Preparing for a Career in Engineering

Bill Fortney
Mark Meno

Preparing for a Career in Engineering

Copyright © 2020 by Bill Fortney and Mark Meno

ISBN 978-1-71747-707-1

All rights reserved. No part of this publication may be reproduced, distributed, or transmitted in any form or by any means, including photocopying, recording, or other electronic or mechanical methods, without the prior written permission of the author. Send permission requests to wbfortney@gmail.com

Cover photo by Pam Fortney
Editing by Michelle Pierce and Pam Fortney

Dedication

We dedicate this book to God - the beginning of all knowledge.

He has blessed us with wonderful careers, and now He is allowing us to pass on to the next generation of engineers.

To God be the glory!

I also dedicate this book to my father, Thomas Guy Fortney, Jr MD. His love and great example taught me many of the lessons I discuss in this book. He demonstrated what it means to work hard and the importance of always giving that little bit of extra effort to ensure the job is done well. He modeled integrity, kindness, humility, and respect for all people. He deeply loved my mother and gave me the precious gift of a loving home to grow up in. Last, but certainly not least, he endured my endless "why" questions and helped me to develop a love of learning. Thank you, Dad. It is an honor to pass on some of your great wisdom.

Bill Fortney

Bill Fortney

I also dedicate this book to my Grandpa and Grandma Meno. As their first grandchild and born on their wedding anniversary, I always felt a special bond with them and, as such, a responsibility to honor them in how I lived my life then and today. As I have grown older and reflect on the years past, memories of them offer me peace and happiness. I use their 60+ years of marriage as a guide. The spirit in which they lived their lives was a benchmark I attempt to emulate. Their faith in God was unqualified and boundless, and it spilled over into everything they did. I miss them.

Mark Meno

Mark Meno

Acknowledgments

A seemingly endless number of individuals contributed content, their stories, constructive input, or their editing skills. Special thanks goes out to: Megan Beszterczei, Angela Bell, Josh Corbett, Alina Creamer, Jonathan Chalfant, Ramsey Davis, Scott Fisher, John Francis, Elijah Hyden, Susan Lloyd, Ali Nejad, Emma Meno, Marissa Olberding, Bob Parys, Michelle Pitman, Nathan Rowland, Sammie Rowland, Andrew Scott, Carli Starnes, Eric Santure, Hannah Teeters, Savannah Tabor, Amy Wheary, Jim Yankauskas, and Amie Zales.

Bill

This book would never have happened without the wise counsel of my wife Pam and the enduring commitment from many of my initial MES students (Cliff, John, Jordan, Josh, Laine, Ramsey, Holly, Marissa, Travis, Andrew, Lesley, Daniel, Kevin, Sam, Brenda, Laura, Jonathan, Cole, Jason, Jason, Caleb, Matt, Jordon, Justin, Dakota, Rodney, Carli, Taka, Brad).

I returned to teaching from industry in 2001. My expectations were high, and I assumed that students not meeting them did not care. This assumption turned out to be far from the truth. My wife taught me to see each student as a complete person and to care about them instead of just their performance. I began to see their deep commitment and their amazing talents. The quality of who they were forced me to look deeper into the performance gap, and it was not due to a lack of effort. It was due to a lack of knowledge. I assumed that students knew things they did not, and I expected them to perform in a way in which they had never seen. I needed to change, and this book reflects many of my lessons. Pam/MES Students, please know I deeply appreciate how you helped me.

Mark

My wife Martina, whose willing pause in her personal career to stay home and raise our two beautiful, smart, and successful daughters (Emma and Cora) enabled me to focus on my career and secure leadership positions that have provided me great career opportunities. Without her, my input to this book is not possible. I love you, Tina.

My Mom and Dad, who made it a single focus of their lives to ensure I was successful in everything I did, often through their own sacrifices. While that caused some "beneficial friction" along the way, it trained me to never stop pursuing perfection. Thank you, Mom and Dad, for pushing.

My first boss, Greg Piner, who took a chance on a kid with a "less than excellent" GPA. To this day, I don't know what he saw in me that led him to not only ensure I was hired but to push me into leadership development programs and to support my rotations in hopes I'd one day be in a position of influence. He even went so far as to hook me up with a couple of the best early career mentors one could ask for – Jack Fennell and James Whitfield. Thanks, Greg.

TABLE OF CONTENTS

Chapter 1 – Preparing for Success 1

Chapter 2 – Essential Skills for Success 9
    Be Prepared to Work Hard 9
    Be an Owner 10
    Learn with the Intention to Act (Fill Your Tool Belt) 12
    Be Organized 14
    Practice Self-Assessment 19
    Manage Personal Growth 20

Chapter 3 – Be Successful in School 29
    Treat School Like a Job 30
    Learn How to Study 30
    Get the Most from Class 33
    Use Your Resources 36
    Next Steps 37

Chapter 4 – Begin With the End in Mind 39
    Accept Responsibility for Solving Technical Problems 42
    Act with Professionalism 43
    Next Steps 46

Chapter 5 – Practice the Discipline of WHAT Before HOW 49

Chapter 6 – Final Thoughts 61

Appendix A: Differences Between School and Work 65

Appendix B: Success Traits of an Engineer 67

About the Authors 78

Chapter 1 – Preparing for Success

You accomplish big dreams with much hard work and by climbing the ladder to success rung by rung.

September 2, 2015 was a typical Eastern North Carolina fall evening. The temperature was in the seventies, and there was a light five to six mile per hour wind. A special group of Marines was conducting Helicopter Rope Suspension Techniques training aboard Camp Lejeune. At approximately 9:00 PM, their CH-53E helicopter experienced a problem and made a hard landing. Staff Sgt. Jonathan Lewis lost his life that day [1]. Whom did they call to understand what happened with the helicopter and to ensure it never happened again? They called an engineer. Specifically, they called a team of engineers, which included one of my (Bill's) former students, Josh. Josh graduated three years earlier from the NC State Mechanical Engineering Systems program, and it was his responsibility to ensure that everyone involved with this helicopter remained safe. His team completed the rigorous task of determining the root cause of the problem. While a permanent solution was being developed, the team implemented a mandatory inspection for every CH-53E in the fleet. Just considering the pilots and crewmembers, I estimate the work from Josh's team impacted close to 1,000 people or more. If I include family members, that number doubles. Their work even affected me. The CH-53E pilot on the front cover of this book is my son-in-law. Every time he leaves to fly, I remember that Josh and his team help to ensure my son-in-law stays safe and comes home to his family. I am thankful for Josh and for how he prepared to meet the challenges of an engineer.

If you are reading this book, then it is likely that you think you want to be an engineer. This means that one day it will be you, instead of Josh, receiving the call to rise to the challenge and positively affect lives with your technical knowledge. Are you ready?

One day it will be you **methodically** planning a course of action and working through all obstacles to solve a problem.

One day it will be you **thoroughly** studying every aspect of the situation, exploring all relevant facts, and understanding all underlying root causes.

One day, it will be you **systematically** integrating all of the information to make a decision and recommend action.

One day, people will look to you for answers. Are you ready?

Amy was sitting at her desk preparing for the busy day ahead when she received the call from her senior engineer. An aircraft had landed in a grassy field during a training mission, and both the grass and aircraft caught fire. The pilots wanted to fly the aircraft back to its home base, but first, an engineer would have to assess the situation and determine if the aircraft was safe to fly. Amy's senior engineer was sending her to make the decision. Amy had recently earned signature authority, which means she could independently release repairs for which she was confident and comfortable.

After some effort, Amy arrived on the scene and was greeted by anxious pilots and high-ranking officials. They all had one thing on their mind, "Can we fly this aircraft home?" and it was up to Amy to give them an answer. An answer of "no" would be very costly and would require a significant investment of time. The aircraft would have to be partially disassembled on-site and then the pieces lifted out.

As Amy looked around, she suddenly felt alone. Sure, there were plenty of people present, but she, alone, was the technical authority. There was no cell phone coverage, so she could not consult with her senior engineer or any other technical person. She, alone, had to make the decision concerning the aircraft's readiness for flight. She imagined the crew as family and asked herself if she would want her family flying out in the aircraft. She **methodically** and **thoroughly** investigated all relevant aspects of the situation, and then **systematically** did what only she could do. She made the decision.

Like Josh, Amy received the call to affect lives with her technical knowledge positively, and she was ready. Are you ready?

The stories of Josh and Amy depict your future as an engineer. Notice two things.

First, notice that their work made a difference. No matter what type of engineer you become or what industry you work in, your work as an engineer will affect lives. At the other end of every engineering decision you make are people, and your decisions directly touch their lives.

Second, notice that people counted on and listened to them. You will be highly respected as an engineer, and people will depend upon you for your technical knowledge. They will look to you for technical guidance and will trust what you say to be correct. Like Josh and Amy, you will often have the final word, and your decision will put many actions into motion. This trust people put in us is why engineers learn to do everything **methodically**, **thoroughly**, and **systematically**. We cannot afford to overlook something or to make a mistake.

So, are you ready to meet the challenges and responsibilities of an engineer? Are you ready to rise to the challenge and positively affect lives with your technical knowledge?

Are you ready? Of course, you are not ready - today. No one expects you to be - yet.

You are at the beginning of a long journey, and no one is ready at the beginning. Some start with a stronger academic background than others do. Some have more life experiences than their classmates do. Nevertheless, no one is ready. You are not ready now, but you can be ready. There was nothing special about Josh when he was where you are now, and he made it. He started with limited technical knowledge, low confidence, moderate leadership skills, and no understanding of what it means to be a prepared engineer. Amy was not ready, either. They both started right where you are, but they developed and were ready to meet the challenge of being an engineer.

> At the other end of every engineering decision you make are people, and your decisions directly touch their lives.

You, too, can be ready, but you need a ladder. It is a long way from where you are to where you need to be, and you will get there by climbing the success ladder one rung at a time. The following true story about a young man from Cleveland, Ohio named JC illustrates this climb [2].

As a young man, people described JC as frail and skinny. When he was 13, Charlie Paddock visited JC's junior high school. Charlie was a great American sprinter and was known as the world's fastest human being. Charlie spoke to JC and his classmates and told them about his experience winning at the 1920 Olympics. Afterward, JC's coach, Charles Riley, invited JC to his office to talk more with the great Paddock.

This encounter with Charlie Paddock planted a seed that would change JC's life. He thought it would be nice to be known as the fastest man in the world, and he told his coach about his dream. There in his office, Coach Riley looked at JC and told him something that he never forgot. He told him that his dream was high and that he would have to climb a ladder to reach it.

JC's coach knew that a person does not accomplish big dreams like becoming the fastest man in the world (or your dream of becoming a prepared engineer) by chance. You accomplish big dreams by making a plan to achieve them. You accomplish big dreams with much hard work and by climbing the ladder to success rung by rung.

JC climbed the ladder, and his hard work paid off. At the 1936 Summer Olympics in Berlin, Germany, JC became the first American in the history of Olympic Track and Field to win four gold medals. You likely know JC as Jesse Owens. Jesse broke the Olympic and World Record in the 200 meters, and he set a broad jump record that lasted for twenty-four years. Jesse rocked the world!

Jesse set his sights on what he wanted, and he climbed the ladder to get there. He did not sit back and hope his dream would come true. Rung by rung, he took steps to make his dream a reality.

Like Jesse, you cannot sit back and hope your dream of being a prepared engineer comes true. I will talk more about this later, but just going to and doing well in your engineering courses will not completely prepare you to perform well as an engineer. The technical knowledge achieved from your classes is one critical rung in your preparedness ladder, but there are many others. To be prepared like Josh and Amy, you must take ownership for understanding and climbing these other rungs. This book will help you make the climb.

Several years ago, I decided I wanted to participate in a sprint triathlon. I was an avid road biker and became inspired by my friend Barry who completed the ironman challenge. When I told Barry about my interest in a triathlon, he offered to help me prepare. I quickly accepted his offer because the thought of preparing for and participating in a triathlon terrified me. It seemed like an unreachable dream, and I had no idea where to start. You may feel this way when you think about preparing to receive the call to action as Josh and Amy did.

Like Jesse Owen's coach, Barry gave me the necessary ladder. He knew exactly where I was going and how to get there. He had gone ahead of me and been successful, so he helped me prepare rung by rung.

Mark and I are like my friend Barry. We have both been down the road of preparing for a career in engineering, and we did it well. You can read more about our careers at the end of the book, but we performed well in our first job and quickly advanced in our companies. We have both hired and managed engineers, and we know firsthand the skills and characteristics required for success in engineering.

Mark has served in many roles within his organization, from the Senior Civilian for the facility to the Head of Research and Engineering, employing hundreds of engineers from dozens of top-rated universities. They support military aviation and influence daily the safety of thousands of people all over the world. I am the head of North Carolina State University's site-based Mechanical Engineering Systems (MES)

BSE program. For over fifteen years, I have been using my experience from industry to help students prepare for the challenges of a productive engineering career.

My career started in industry and transitioned into education. Mark is still in industry, but he demonstrates an educator's spirit. Together, we have insights to help guide you in your journey to preparedness. If you let us, Mark and I want to be your guides from now until your graduation. We want to help you develop the same **methodical, thorough,** and **systematic** approach you saw in Josh and Amy. We want to help you develop the character traits that will cause people to trust you with great responsibility.

To climb a ladder safely, you start at the bottom and go up one rung at a time. Barry helped me do this. There were many decisions to make for success on race day. A few are below.

- Where will I start in the open water swim? Will I start behind the pack, to the side of the pack, or in the middle of the pack?
- How and when will I fuel? Will I use liquid or solid fuel?
- What will I wear? Will I change after swimming or bike and run in my wet clothes?
- What pace will I start with on the bike and the run?

Making these decisions was crucial to my preparedness, but Barry did not mention these to me. As a wise guide, he knew these decisions were for higher rungs and would overwhelm me at the beginning. Instead, he started with where I was and helped me to build the basic skills and endurance needed for success. As time went on and my knowledge and experience grew, he added rungs, and I worked on other aspects of success. Eventually, he brought up all of the decisions I needed to make to be prepared for race day, and I easily made them. At the bottom of the ladder, these decisions would have overwhelmed and discouraged me. After climbing the ladder and preparing rung by rung, I confidently made them.

Mark and I understand that you have a big task ahead of you and that the only way forward is to climb rung by rung. As Barry started with the basics and then built on them, we intend to do the same. This book provides the fundamental thinking and skills you need throughout your journey of completing school and entering the engineering workforce. You will come back to this book often and work on the skills throughout your time in college. You can even use it as you traverse through your career.

When preparing for my triathlon, I had to learn patience and pace myself. If I tried to progress too quickly, there would likely be an injury, and this would be counterproductive to my progress towards preparedness. You will need to be patient and pace yourself as well. You will not just sit down and read this book through as you would a typical book. You may want to read it through once, but then you will go back, read a chapter, and spend some time reflecting on the material before you move on to the next chapter.

Reflection may be a new idea to you, but it is simply taking time to give serious thought. Personally, I find reflection very difficult, and initially, it may be hard for you. I naturally focus on the current task, complete it, and then immediately move on to focus on the next task. Stopping and giving serious thought is not natural for me, but I have learned that it is essential to growth.

To help you develop good reflection skills and pace yourself throughout the book, Mark and I created reflection questions for each chapter. These questions will help you stop and think about the key concepts from the chapter. Our hope is that you will stop after each chapter and spend some time thinking about the material by using the chapter's reflection questions. Like my triathlon training, getting in a hurry will be counterproductive, so take your time and internalize the material from each chapter.

Throughout the book, you will see scrolls like the one below. These are Peer Insights, and they contain comments from current engineering students and recent graduates who reviewed this book. We hope their comments help you understand the relevance of what you are reading and inspire you to press on in your journey.

> *I really appreciate that you used so many references to "real-world" experiences. As students, we tend to think of what comes after school, or even events that happen during school but are seemingly unrelated to our studies, as an entirely different world. Because of this, oftentimes we don't understand the gravity and significance of the material we are learning.* Alina

I followed Barry's guidance and allowed him to lead me on a six-month journey to prepare for my triathlon. I will be honest with you, it was hard, and I often wanted to quit. He provided a plan, but I had to follow it and put in the work. I spent many lonely hours running, swimming, and biking, but race day came, and I had a blast. I did not place (people in my age bracket are very serious about these events), but I met every performance target I set for myself and left very satisfied. Due to the hard work and step-by-step preparation, the thing that once terrified me became a source of great enjoyment.

Like preparing for my triathlon, there is no magic formula for being a prepared engineer. This book shares some insights with you, but it is up to you to study them, practice them, adapt them, add to them, and allow them to transform you.

There will be many lonely hours studying, and you may often want to quit. Please do not. Put forth the work, follow Mark and me on this journey, and you will find great joy at the end. When you receive the call to action as Josh and Amy did, you will be ready and will experience the joy of making a difference.

Are you ready? Are you ready to start preparing? Read the next chapter and begin climbing.

CHAPTER 1
REFLECTION QUESTIONS

Remember, these questions are here to help you stop and think about the key concepts from the chapter. Take your time and reflect on how you can internalize the material you just read. For maximum benefit, get a composition notebook and write your thoughts to each question. Writing can help you process information, and it allows you to go back and review your thoughts later.

1. How did you feel after reading the stories of Josh and Amy? What thoughts did you have?

2. What are the three words used to describe how Josh and Amy approached their tasks? Why do you think these words were highlighted in the book?

3. What do you think about the statement that just going to and doing well in your engineering courses will not completely prepare you to perform well as an engineer?

4. What do you think about the concept of the ladder from the story of Jesse Owen?

5. What are Peer Insight boxes?

6. Does this chapter leave you excited or scared about being an engineer?

7. What thoughts do you have about your journey ahead?

8. Are you ready to start preparing to be an engineer?

Chapter 2 – Essential Skills for Success

I have seen many students like Josh develop and become very successful engineers, and they all have one thing in common. They felt accountable for the outcome of their education and took ownership for being ready.

Recall the story of my friend Barry and how he helped me prepare for a sprint triathlon. One of the first things Barry did was to help me learn about nutrition and recovery. Proper knowledge of these topics was essential because they are the foundation on which the rest of the training builds. Until I learned to practice good nutrition and recovery, my training would be ineffective and maybe even counterproductive. The topics in this section are your foundational skills on which everything else will build. Understanding and practicing these skills is what will allow you to build other skills necessary to become a prepared engineer. You will use and continue to develop these skills throughout school and your engineering career, so make it a habit to read this chapter often.

Be Prepared to Work Hard

Becoming an engineer is a great dream. Together Mark and I have been engineers for over fifty years, and we can tell you that it is a wonderful field. Engineering gives you the foundation to pursue a wide range of careers and allows you to perform work that makes a difference. Engineering is an extremely fulfilling career, but I will not mislead you. Like any great thing, the journey to becoming an engineer is hard. Fortune Magazine describes a study that looked at why certain people excel in their field [3]. The researchers studied leaders in various fields, such as sports and business, to determine if they were just born great or if something else made them excel. The conclusion was simple. The common factor in all people who excel in their field is HARD WORK! An excerpt from the article is given below.

> "The best people in any field are those who devote the most hours to what the researchers call **'deliberate practice.'** It's **activity**
>
>   that's **explicitly** intended to **improve** performance,
>
>   that **reaches** for objectives **just beyond** one's level of competence,
>
>   provides feedback on results and involves **high levels of repetition**."

Do not miss the conclusion from this study. You cannot accomplish something great like becoming an engineer by just showing up and going to class. Good engineers like Josh and Amy are good because they work hard and are intentional about developing what is necessary for success. They were intentional and worked hard in school when preparing, and they continue to work hard every day on the job and are intentional about learning as much as they can about their field.

Notice the phrase from the study "deliberate practice." You will hear Mark and I use this phrase often, and when we do, your mind should go back to this study. Many of the skills necessary to become a prepared engineer will not come easy, and you will need to be intentional and use deliberate practice as described above to build them. At the end of this chapter, we discuss a process that will help you intentionally develop these skills for success.

> You cannot accomplish something great like becoming an engineer by just showing up and going to class.

Be an Owner

If you have ever rented a car or house, then you know that you look at it differently than you do your own car or house. As the owner, you feel completely responsible for everything about the car or house because it is yours. As a renter, you do not feel this same responsibility. You feel like someone else is responsible. Preparedness requires being an owner of your education.

In his blog, FridayForward, management expert Robert Glazer describes people with an ownership mentality as those that "focused on what they could control but also took responsibility for those external variables" [4]. In his company's Core Values, he defines owning it as follows.

> "Owning it means being proactive and taking accountability for outcomes, even when variables are beyond our control and ambiguity is present" [4].

Do not miss the word accountable because it is key to understanding ownership. I have seen many students like Josh develop and become very successful engineers, and they all have this one thing in common. They felt accountable for the outcome of their education and took ownership for being ready.

If you want to be prepared, then accept the fact that you, alone, are accountable for your preparedness. It is not up to your teacher, advisor, your manager, or anyone else. You are responsible for your preparedness. There are many people along the way to help you, but your preparedness is not their responsibility. It is yours. Being an owner means that you take responsibility for your preparedness and put it completely on yourself.

Also, do not miss the phrase "beyond our control." When problems occur, renters blame and make excuses. Instead of taking responsibility in a situation, they tend to shift the focus to others with phrases such as "they need to," "they did not," or "I cannot because they." Owners accept responsibility and find a way to resolve the problem. It does not matter who caused the problem. Owners accept responsibility and fix it.

> *I find that my fellow classmates who complain the most about instructors and do nothing to resolve the issue end up transferring the same attitude into their place of work. They would rather complain about decisions made by the boss/supervisor with their coworkers than confronting the issue directly in order to solve it.* Elijah

Adopt an owner's mentality towards your education. The examples below illustrate some ways to be an owner in school.

> If you have problems learning from a teacher, blaming them and complaining about how you cannot learn from them will not help you learn. That is being a renter, and you will carry this habit into the workplace. Instead, be a responsible owner and find a way to learn the material. In Chapter 3, we provide a few suggestions on how to do this.

> Do not expect your instructor to tell you everything you need to do to learn the course material. This is having a renter's mentality, and it is like expecting your engineering manager to tell you everything you need to do to solve a problem at work. We will discuss this more in Chapter 3, but you are hired so that some of the workload can be taken from your engineering manager's shoulders and placed on your shoulders. If you wait around for your manager to tell you everything needed to solve the problem, it defeats the original purpose for hiring you. Learn to be an owner by being proactive and using critical thinking to determine what you need to do to be successful in your courses. Before the first class, study the syllabus and carefully review any other course resources such as a learning management system. Owners take an active role in learning. Renters are passive and only act when told.

> If you are having a problem while in school, do not be a renter and complain, make excuses, and blame others. Instead, take ownership for your actions and find solutions. Being an owner does not mean you have to solve your

problems without help. Your instructor or school administrators will gladly help you, but they want you to do all you can to find solutions before you come to them for help. You also have resources such as tutors, others in the class, and people who have already taken the class. Owners take responsibility for learning and utilize all available resources.

Learn with the Intention to Act (Fill Your Tool Belt)

A bad habit that is easy to develop in school is learning something only when there is a homework assignment or when it is going to be on an exam. For example, if the material from this chapter is going to be on an upcoming exam, you will read it and take note of the items you need for the exam. If you have a homework assignment to complete with the reading, you will read the material well enough to complete the homework assignment. If there is no exam or homework tied to this chapter, it is very likely that you will not even read the chapter. You will not read it because there is nothing assigned.

If you are not careful, school will teach you that you learn to complete an assignment or perform on an exam, but this is a dangerous view of learning, and you will carry it into the workplace. As engineers, we do not learn to be able to perform to the expectations of an exam or homework assignment. We learn to be able to meet the expectations of the person looking to us for answers. We learn because people are counting on us to be right. We learn so that we can put the knowledge into practice and **methodologically**, **thoroughly**, and **systematically** develop technical solutions. We learn even when we do not have a specific problem to solve because we know the knowledge will give us skills and allow us to bring more to the next task.

Imagine you are in your living room reading about how to deliver a baby. You likely would learn facts and scan over a few details. Now imagine you are in the middle of nowhere with a woman going into labor, and you are reading about how to deliver a baby. Your reading is no longer theoretical, and your entire attitude would be different. You are now reading with the intention to act. This is the attitude you must have about learning in your courses if you are going to be a prepared engineer.

> If you are not careful, school will teach you that you learn to complete an assignment or perform on an exam, but this is a dangerous view of learning and you will carry it into the workplace.

Have you ever watched an artisan at work? It does not matter the field – a mechanic, carpenter, electrician, or welder. For each, a key to their excellent work is the correct tools in their tool belt. Having the correct tool makes a complicated job look simple. Having the wrong tool can make a mess quickly. Imagine a finish carpenter trying to drive a nail with a screwdriver. It may work, but it will not do the job well. Having the correct tool for the job allows the artisan to perform excellent work.

> Often you hear people say that they never used what they learned in school. This statement is not true for engineers like Josh and Amy that came out of school prepared.

Having the correct tool is essential for an artisan, and it is essential for you as an engineer as well. Your manager will give you a specific problem to solve, and you will need specific personal and technical skills in your tool belt to find the solution. If you have the tools, you can successfully solve the problem. If you do not, your manager will find someone else with the tools, and they will solve it instead of you.

Often you hear people say that they never used what they learned in school. What these people usually mean is that they did not fill their tool belt while in school. They took classes and did work, but they did not learn with the intent to act. Once on the job and a tool was needed, they went to their belt, but it was empty. They concluded that school did not teach them what they needed. In reality, they learned/memorized the material for the exam instead of learning with the intent to act and storing the knowledge in their tool belt.

> *At the beginning of our schooling, my peers and I struggled to understand how much our studies connected to the "real world" and what lies beyond college. When classes got difficult, we shared thoughts like these:*
> *"This may be difficult, but at least we'll be done with it in a semester!"*
> *"I'll never use this out in the real world."*
> *"Understanding the material enough to pass the exams and get the degree is good enough. I'm sure the rest will be taught on the job."*
> *We were wrong.* Alina

To be prepared like Josh and Amy, you must fill your tool belt with personal and technical skills that you are ready to use. The better you learn and prepare now, the better you can perform on the job. Prepared engineers know and can use the information from their engineering classes. Realistically, you cannot remember every detail about every technique you learn, but you should remember that the technique exists and understand how it can help you. The technique is in your tool belt, and the fuller your belt, the better you can solve problems that no one else can solve. A former student told me about a problem he was solving which dealt with an underwater guidance system. There was a problem with drift due to extreme currents, and it was proving to be quite a challenge. The student remembered a technique from a class and, after some review, was able to use it to solve the drift problem. A full tool belt allowed him to solve a problem that no one else was able to solve, and it will do the same for you if you build the habit of learning with the intent to act.

Elijah is a senior mechanical engineering student who works part-time as an engineering intern. He offers these insights on how material from your courses applies on the job.

> Just because someone may not currently use 100% of the knowledge from school, no one knows where they will be several years from now. Engineering is a very diverse field, and it would be surprising to work in a place that uses 100% of the knowledge acquired from school. It would be foolish to neglect a fundamental principle taught in school just because you may not currently use it. I was having a conversation with a coworker, and I was talking about the difficulties of working and going to school. I claimed that it was hard going to a class that I knew I was never going to use again. He replied by asking me, "Do you know where you are going to be in five, ten, or fifteen years?" As soon as he asked the question, I knew what he was getting at, and it dawned on me that he was right. I smiled and laughed to myself as I answered, "No, I guess not." His rhetorical question has stuck with me since then, and I have tried to have a new perspective on any of my classes that I may not currently use at work.

Let me bring this discussion back to this chapter and the entire book. Do not read to perform on an exam or to complete homework questions. Read to fill your tool belt. Read knowing that you must put what you read into practice. If you want to be a prepared engineer, you will need to act on some material you read immediately. Be courageous enough to put what you learn into practice today. With other material in the book, you will learn and store it in your tool belt. When the call to action comes, you WILL pull out the right tool and make a difference like Josh and Amy did.

Be Organized

Your time in school is busy. You always have many tasks to complete at once, and information constantly bombards you. I wish I could tell you that it will get better when you go to work, but it will not. You are balancing many classes now. You will be balancing many projects at work. One of the most important skills you can develop to perform well in school and on the job is organization. Good organizational skills are fundamental to success. You specifically need to develop the ability to organize information and manage your time. These two topics are discussed below along with some organizational suggestions for each. These suggestions are offered only as examples. We all organize differently, so you will want to find what works for you. The exact technique is not important, but it is important to learn some methods to organize information and manage your time.

Organize Information

Beginning with school, engineers are flooded with an overwhelming amount of information. Information comes pouring in from your courses as well as avenues such as emails, texts, peers, the internet, and learning management systems. Your success in school, as well as on the job, largely depends upon your ability to successfully capture, filter, and appropriately use information as illustrated in Figure 1.

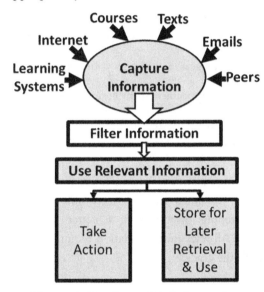

Figure 1: Dealing with Information

Capture. As an engineer and as a student, you cannot afford to miss any relevant piece of information. Missing one key fact can mean failure, so you must learn how to capture or "hear" all information coming from every source.

For information coming from verbal sources such as your instructor, a conversation, or a meeting, this means staying engaged and hearing everything that is said. Active listening and taking notes will help you not miss facts from verbal communications, and we discuss both in Chapter 3.

For written sources of information such as email, you have to 1) regularly check your email, and 2) break the habit of just skimming and learn to read the material in detail. Your instructors (and your manager) will often send critical information to you through email, and they expect you to read the email in a timely manner and not miss any detail given. I find that your worst enemy in this area is focus. Most engineering students can be single minded and focus on one thing for long periods of time. This skill is great for doing well in one hard class. It is a disaster for keeping up with all of the other details of life that continue to bombard you.

> It is important to learn how to periodically break focus and catch up on details such as emails concerning other courses.

Students often focus so much on one thing that they miss everything else. It is important to learn how to periodically break focus and catch up on details such as emails concerning other courses. To keep email information from falling through the cracks, consider the suggestions below.

- Have a set time each day to carefully go through all emails and identify any important information. "Carefully" means you need to read and not skim. This is the only way to ensure you do not miss key details.
- If the item needs action, add it to your running to-do list.
- If the item is an appointment or assignment, add it to your calendar.
- If the item is information you need to refer back to later, capture it as described below under Use – Store and Retrieve.

Filter. Ensuring you capture all of the information coming your way is only the first step. After capturing it, you must learn to filter out the noise from the relevant information. Active listening discussed in Chapter 3 will help you filter the information you take in. All relevant information requires attention, and you cannot let any of it fall through the cracks.

Use - Act. Some relevant information requires you to take action and do something. For example, a due date given in an email, a homework problem assigned during class, or an email announcement concerning a scholarship opportunity. Once captured, this type of information is easy to deal with if you learn good time management techniques.

Use – Store & Retrieve. You will not immediately use most of the relevant information you take in. Learning to manage this information will require the most work. For example, one of your engineering professors gives a hand out during class that explains a certain analysis method. The method will not be on your upcoming exam, but it is likely you will need to use the method later in the semester or in a future course.

You first need to have a way to store the material on the analysis method so you can find it when needed. Stuffing it in your backpack is not a way. One organizational method is to have a notebook or electronic folder for each class with divisions such as class notes, reading notes, homework, and handouts. Find a method that works for you, and use it so you can locate information when needed.

You also need a way to remember that the material exists when it is time to use it. Mark calls this a mental filing system. Recall the story of the student having trouble with drift due to strong currents when trying to design an underwater guidance system. The student used his mental filing system and remembered that a technique existed which might help with the current problem. This is exactly what you want to learn to do. You will not remember all of the details for every concept and technique you encounter in your courses, but you should file enough information to be able to identify when a stored concept or technique is relevant to your current task.

> Developing some type of mental filing system is critical, because it is how you as an engineer can be thorough and consider all relevant areas when solving a problem.
>
> File enough information to be able to identify when a stored concept or technique is relevant to your current task.

Developing some type of mental filing system is critical because it is how you, as an engineer, can be **thorough** and consider all relevant areas when solving a problem. You **systematically** move through your mental filing system and identify relevant concepts and techniques that you have stored from your courses and other study. You then **methodically** solve the problem. To help develop your mental filing system, always look for ways to apply knowledge from one course in other courses.

Managing Your Time

As stated earlier, you are busy as an engineering student, and you will continue to be busy at work. It is essential that you learn to manage your time well. There are many good internet resources on time management, and I encourage you to find one that works for you. I offer a few tips below.

- Get a planner or electronic calendar and use it.
- Create a Weekly Schedule.

- For a given semester, your class schedule is fixed, and it can be very helpful to create a weekly schedule like the example shown in Figure 2. You can use your electronic calendar as long as you add the detailed time information for each class to it.
- A weekly schedule gives you a quick snapshot of your "open" times, and you can use these to schedule study time and team meetings. Yes, "schedule" study times.

	Mon	Tues	Wed	Thurs	Fri
8:00	Calc-II		Calc-II		Calc-II
9:00					
10:00		Physics			
11:00	Phy-I	Lab	Phy-I		Phy-I
12:00	Walking		Walking		Walking
1:00					
2:00					
3:00		ENG		ENG	
4:00					
5:00					
6:00					
7:00	work	work		work	work
8:00					
9:00					
10:00					

Figure 2: Example Weekly Schedule

- In most jobs, you work eight to nine hours, five days a week. Think of school the same way. I call this adopting a 9 to 5 mentality. View your time at school like time at work. During "work" hours, you work. On Monday, for instance, do not waste the 2 hours between Calculus-II and Physics-I. After Calculus-II, take a few minutes to clear your head, and then get back to work until it is time to go to Physics-I. You will be amazed at how much you can accomplish if you do this.

> *It's amazing how much can get done if you maintain the "work" mindset, and in turn, it keeps you on track for your afternoon classes. This also transfers well to the workplace, where you can't stop working for several hours in the middle of the day.*
> <u>Nathan</u>

- Go through each course syllabus and add the due dates for all assignments, projects, and exams to your planner or calendar.
- Use a running to-do list. When something comes up that you need to do, add it to this list. An item stays on the list until it is completed. Some people like to have a list of long-term items and another for short-term items.

- Once a week (I like to do it on the weekend), do the following.
    - If you have emails you have not dealt with as described above, do this now.
    - Look at your planner or calendar and review the next two weeks. Identify what is due (assignments, projects, exams) and then schedule blocks of time to work on each activity. The weekly calendar (Figure 2) comes in handy here since it easily shows your "free" blocks where you can schedule some time to work on assignments.
    - Review your running to-do list and see if there are items you want to schedule to work on during the coming week.

If you do not do some planning, you will find yourself in constant crisis mode. Monday morning, you will realize that you have three exams, two projects, and three homework assignments all due that week. You know what happens next. Learn to plan, so this does not happen.

> You specifically need to develop the ability to organize information and manage your time.

Practice Self-Assessment

If you take a realistic look at yourself, you know that you are not where you need to be with the skills for success in engineering. You have some improvements to make. Do not feel bad because this will always be the case. No matter how much experience you have, there will always be room to grow and improve. Professionals are always assessing their performance and improving.

In school, your instructor constantly gives you feedback on how you are doing through homework and tests, but at work, it is going to be you who assesses how you are doing. Sure, you will get feedback from your manager, but this is typically very infrequently unless you really mess up. No news does not always mean good news at work. Often, by the time you hear from your manager, poor performance has already damaged your reputation. Your peers have heard about the poor work you do and will avoid working with you. The same will happen with your peers in school. To keep this from happening to you, don't wait for others to tell you how you are performing. Instead, take ownership for your performance and push yourself to excel by learning to practice the skill of self-assessment.

The discussion on what it takes to be great from earlier in this chapter is a description of self-assessment. Self-assessment is determining where you need to improve by deliberately taking an honest look at where you are and where you need to be. You can fool everyone else, but if you are honest, you will never fool yourself. You know the kind of effort you are giving and the quality of work you are producing. To be a successful engineer, you need to develop the habit of honestly assessing your performance and making the appropriate adjustments to your behavior.

Self-assessment begins with an honest look at where you are. This look is often called reflection, which simply describes taking time to give serious thought. I mentioned in Chapter 1 that reflection is very difficult for me, and it may be for you as well. Stopping and giving serious thought to my actions and the past does not seem natural, but I have learned that it is essential. You cannot improve unless you have a realistic look at how you are currently doing, and reflection gives you this look. I had to exercise deliberate practice to learn the skill of self-assessment through honest reflection, and it is likely that you will have to do the same. To practice, you can reflect and assess yourself after an event such as a homework assignment, exam, or team meeting. Another option is to reflect weekly or monthly. I recommend at least setting aside a specific time each month to reflect and review how you are doing. The method discussed at the end of this chapter will help you practice self-assessment.

> Don't wait for others to tell you how you are performing. Instead, take ownership for your performance and push yourself to excel by learning to practice the skill of self-assessment.

Manage Personal Growth

The book "The Boys in the Boat" tells the extraordinary story of nine ordinary men from the University of Washington who seemingly come from nowhere to win the gold medal in rowing at the historic 1936 Berlin Olympics [5]. Against financial, personal, political, and physical odds, they emerge as winners. They should not have been able to win, but they did. Several years ago, my wife and I visited the University of Washington. As we approached the rowing facilities, I noticed banners on all of the light poles that read, "Who We Are is Why We Win." When I saw this slogan, I thought that it precisely described the essence of engineering success. Technical knowledge (what you know), alone, does not make you a good engineer. You will excel as an engineer because of who you are. Notice that the items in this chapter mainly deal with personal traits and attitudes. It has focused on topics associated with who you are. Becoming a prepared engineer means developing what you know and who you are. This means change.

Change is hard for all of us, but it is especially hard when dealing with personal traits and attitudes. I wish I could give you some magic formula to make change easy, but there is not one. Recall the idea of deliberate practice. The best people in every field are the best because they deliberately identify where they need to improve and then intentionally work on the identified weak areas through high levels of repetition. We will help you identify areas for change through self-assessments, but you will need to take action and exercise the deliberate practice necessary for change.

The six general steps in Figure 3 form an improvement cycle that will help you intentionally identify areas for improvement and develop the skills necessary to be a prepared engineer. For any Star Wars fans, we call this cycle I2-R2 after the droid R2-D2 (Artoo-Detoo). Take a few minutes and carefully study each of the six steps.

Identify specific areas to work on

1. Read the material in this book with an intention to act, and write down specific things you need to work on. This is your <u>running list of improvement areas</u>. Every time you read, reflect, or perform self-assessment, add to this list. You may also identify improvement areas from classes, seminars, extracurricular events, or a friend.

2. Select two or three things from your list to work on and write them on a clean sheet of paper. These are your <u>active improvement areas</u>.

Plan ways to **Improve**

3. For each active improvement area, decide the steps you want to take to bring change. Be specific such as "I want to actively listen and take notes in one class for the rest of the semester," or "I want to get involved in Toastmasters to practice public speaking." Always keep in mind the "why" behind the item you want to work on. If you are not sure what to do to improve, ask an instructor or go to the career center for help. <u>Write down the actions you want to take under each of your active improvement areas</u>.

Review your progress

4. Tell a trusted friend what you are working on, and ask them to check with you periodically to see how it is going. This friend is your accountability partner, and their job is to encourage you in your journey to improve. Give them a copy of your active improvement areas with your action steps listed under each area. A fellow classmate works well for this accountability partner since you will see them often.

5. Practice self-assessment and monitor your progress.

Repeat and work on something else

6. When you are comfortable with your progress on your active improvement areas, review your running list of improvement areas and start working on another item.

Figure 3: I2-R2 Improvement Cycle

As you read, this book instructs you to review the material, reflect, and practice I2-R2. Please do it. Deliberate action is the only way you will develop into a prepared engineer. At first, it may be difficult going through these steps, and that is understandable. You are changing habits and patterns of thought that have been with you for a long time. Changing habits is no easy task, so I offer some insights based on experience working through change in my life.

> You don't know exactly what to do to improve, so you do nothing.
>> We all like to be right, but engineers especially like to be right, and this desire can work against us. You see an area that needs improvement, but you cannot decide the exact best way to achieve improvement. You research and talk to others, but still, you are not sure. The uncertainty causes you to do nothing because you want to do the "right" thing. I look at the situation as follows. Anything you do is an improvement over doing nothing, so don't let your fear of not getting it exactly right stop you from acting. Perform your research, talk with people, and then do something.
>
> You are afraid of failure, so you do nothing.
>> Engineers like to fail even less than we like to be wrong, but failure is a normal part of engineering, and you must get used to it. When trying to change habits, you will fail. We all do, so accept this as a fact right now. When you do fail, be thankful for the little progress you made before failing and then keep working on the change. Imagine you are trying to build the habit of going to class prepared. For two weeks, you go prepared for every class, but then you fall back into the old habit of not preparing. You may look at this as a failure. I view it as a success. You just went to class prepared more times than you did the previous semester. This is good, and you should celebrate it.

Changing habits is hard, and it will not happen without deliberate action. You will always take a few steps forward and then some back. Celebrate little successes. You have a long journey ahead of you, so pace yourself and climb one rung at a time. Don't get discouraged when you fall short of your expectations, and do not quit!

I once heard the concept of making vector changes in our lives. To understand this concept, think of an airplane flying from Sacramento to Boston. The airplane can take off from Sacramento and follow a straight-line vector to Boston. Notice the top arrow (vector) in Figure 4. Imagine the result if while taking off the airplane makes a slight change in its straight-line course. With just a slight change at the beginning of the journey, it will follow the bottom vector and end up in Washington, D.C. instead of in Boston. A small change at the beginning caused the aircraft to end up in a very different place.

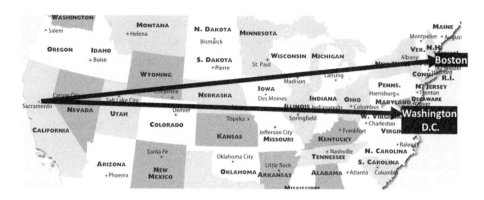

Figure 4: Illustration of a Vector Change

Our lives are like this plane. A few small "vector" changes now can drastically alter where we are in the future. Do not try to change everything at once and end up discouraged and accomplishing nothing. Instead, change a few things at a time. Remember, you are at the beginning of a long journey to becoming an engineer, so pace yourself. Focus on using I2-R2 to develop a few traits and habits at a time. If you diligently practice I2-R2 over the next few years, you will find that 1) you have many of the necessary traits of a prepared engineer, and 2) you have a habit of self-assessment and improvement that will propel your career.

Remember, doing something is better than doing nothing, and we all face failure while improving. Start by developing the essential skills from this chapter. It is going to be hard, and success begins by accepting ownership for your preparedness.

You can be prepared, so begin with I2-R2 today. The reflection questions on the next page will help you get started.

CHAPTER 2
REFLECTION QUESTIONS

This chapter covered much critical information, so there are many reflection questions. These concepts are fundamental, so do not be in a hurry to move on to the next chapter. Take your time and work through the reflection questions below. Write your thoughts in your composition book. As you reflect, remember to add items to your running list of improvement areas.

Be Prepared to Work Hard
1. What is deliberate practice, and how can you use this concept to prepare for a career in engineering?

2. Think about your current work ethic in your courses. Are you happy with it? Do you work as hard as necessary to excel in your courses, or do you just do enough to get by? What changes, if any, do you want to make in this area?

Be an Owner
3. In the context of school, what is the difference between a renter and an owner?

4. What do you think about the idea that you must accept ownership for your performance in a course rather than relying completely on the instructor?

5. Think about your attitude towards your classes. Have you been a renter or an owner? Have you felt it was the teacher's job to teach you (you are just a renter there to do what they tell you), or have you taken ownership and initiative like you are responsible for learning and getting the most you can out of your courses (owner responsible for the outcome)? Have you taken the initiative to get any help you needed to be successful this semester? Did you look in the syllabus and in the course learning management system to see what was expected of you, or did you want someone to tell you what you needed to do? Where do you need to grow in your attitude towards your courses?

6. How are you doing with making excuses (in your classes and projects)? Have you made excuses or kept the focus on completing the task and feeling you are responsible for completing it? Are there things that you could have changed for the good if you had taken action instead of complaining and making excuses? Is there anything you want to do differently?

7. What do you think about Elijah's comment about transferring the habit of complaining to your place of work?

Learn with the Intention to Act (Fill Your Tool Belt)
8. What is the difference between how learning can be viewed in school (the common bad habit of learning in school) and the way the book says learning should be viewed for an engineer?

9. Can you relate to the habit the book refers to of only paying attention to the material if there is an assignment or it will be on an exam?

10. What is the point the book makes with the story of delivering a baby?

11. What does the book mean that you need to fill your tool belt while in school?

12. What do you think about the book's response to the common statement, "I never used what I learned in school?"

13. Think about your attitude towards your courses this semester. Are you learning with the conviction that you will need to use the material someday or just getting through a course?

Be Organized
14. The book says that you will have just as many demands on your time when you go to work as you do now. What do you think about this?

15. How do you think you do at capturing all relevant pieces of information coming at you from the various sources?

16. What do you think about the idea that focus can be your worst enemy? Can you think of times where you focused on one thing and missed details from others? What happened?

17. Do you ever miss important details because you skim instead of read things like email?

18. Do you have a method to ensure you do not miss key information from emails? What do you think about the method presented in the book?

19. How are you doing with storing material you receive from your classes? Do you have a method to store it in an organized way?

20. What is the idea of a "mental filing system," and why does the book say it is so important? What can you do to better store information from your courses?

21. Do you deliberately look for ways to apply concepts you learn in one course in other courses? What do you think about trying to do this?

22. Do you use some type of planner or calendar? If not, do you think using one could help you manage your time?

23. What are your thoughts on the concept of adopting a 9 to 5 mentality?

24. How do you spend your time when not in class? Spend a few days noting how you spend your time. You may be surprised. Don't let things rob time from you. Instead, decide how much time you want to spend on various activities and discipline yourself.

25. Do you use a running to-do list? If not, do you think it would help you to start?

26. What do you think about the idea of having a weekly planning session?

Self-Assessment

27. What does the book mean that no news is not always good news?

28. What is self-assessment, and why is it so important to becoming a prepared engineer?

29. Like Bill, will it take work for you to learn to stop and reflect?

Manage Personal Growth

30. What does the book mean that you have to develop "who you are?"

31. What comes to mind when the book talks about how becoming a prepared engineer means change?

32. What do you think about using something like the I2-R2 Improvement Cycle to help you to be intentional about developing yourself?

33. Which step in the I2-R2 Improvement Cycle do you think will be the hardest for you?

34. What are your thoughts on having an accountability partner to help you work on specific areas of growth?

35. How do you relate to the two common problems to avoid when seeking to change?

36. How does the concept of vector changes relate to you and your journey to becoming an engineer?

37. What do you plan on doing with the challenge to begin intentionally building the personal traits that will make you a successful engineer?

Chapter 3 – Be Successful in School

People will expect you to know and be able to use the information from your engineering classes. They will expect your tool belt to be full, and school is where you fill it.

One essential skill to being as prepared as Josh and Amy were is technical knowledge. Recall the discussion on how to view learning from the previous chapter. People will expect you to know and be able to use the information from your engineering classes. They will expect your tool belt to be full, and school is where you fill it. The fuller your belt, the better you can solve problems.

You can make a good grade in a class and not be prepared to use the material from the course, so view each class as an opportunity to prepare yourself. Study to master the material from your courses instead of seeing your courses as something to just complete. In engineering, there are no optional topics. Beginning with Algebra, all of your courses will build upon each other. Master each topic, because you will see it again.

I remember talking with two students who were both taking Dynamics. They often studied together, but one was making an A in the class, and the other was making a D. I asked them about the difference between their grades, and the "D" student immediately spoke. He said that often he would look at the "A" student and ask her how she knew to use a certain technique to solve a problem. She would reply with, "Don't you remember from Calculus II." He told me that he did not remember and that his lack of knowledge was the problem. Poor study skills and work ethic in a course a year ago were causing him to perform poorly now. The student ended up performing satisfactorily in the course, but he had to relearn Calculus II while he was trying to learn Dynamics. Master each topic the first time so this does not happen to you.

> In Engineering, there are no optional topics. Beginning with Algebra, all of your courses will build upon each other.

At this stage in your journey toward preparedness, learning to do well in your courses is the top priority. Read the following tips with the intent to act and put them into practice.

Treat School Like a Job

One simple thing you can do now to prepare for your engineering career is to treat school like a job. When discussing managing your time, we talked about viewing school like a job with the idea of adopting a 9 to 5 mentality. Treating school like a job also relates to your performance. You cannot act one way in school and then just walk on the job and act differently. Start now acting as if you are on the job. Perform for your instructors the way you want to perform for your manager.

> *Treating school like a job is the best advice any incoming freshman can receive. This mentality rings especially true to me because of the scholarship program I received in school. I absolutely treated school like a job. Transitioning into the workforce felt to me like transitioning from one job to another rather than graduating and beginning a job. That mentality helped not only while I was in college, but has helped and is helping as I begin my engineering career.* Jonathan

Learn How to Study

To fill your tool belt, you must learn to study. When I say this, many engineering students immediately think that this does not apply to them. They only think this because they have not yet taken an engineering class. I speak with many years of experience when I say that their attitude changes. Listen as Mark tells of his first engineering course.

> I was a good student in high school - graduated in the top 5% of my class and took all the college preparation classes I could get my hands on. I was a busy student as well. I was a three-sport athlete all four years, a member of the marching, concert and jazz bands, and an active participant in clubs and school drama productions. My study habits (learned in and around that busy high school agenda) were adequate and effective for meeting high school requirements, so I thought I was ready for whatever college had for me.
>
> I was wrong and had no idea of how unprepared I was for my first engineering course. I soon would find out.
>
> My high school did not have engineering classes, so my first one was as a freshman in college, and we had a test about a month into the fall

semester. I studied (in the only way I knew how). I was prepared (or so I thought). The test was really hard, but I thought I did OK.

I received the graded test back a few days later. It was an 18. Not out of 20 or 25. It was 18 out of 100. This grade shocked me. I was smart and a good student. A top performer. Nothing like this had ever happened before. I was used to succeeding at what I did, so failure was not something I had experienced. It felt like someone punched me in the gut.

It took me a few days to get over the shock of failure. There were many thoughts and emotions, but ultimately, I had to be honest with myself and face my own inadequacy. I did not have the study skills I needed to be successful. It was a hard but necessary admission.

At this point, I really had two choices. I could quit engineering, or I could learn to study.

About 1/3 of my large freshman class chose to quit. Just like that, 30 of my peers chose to change their major out of engineering. I cannot help but believe that many of them would have been very fulfilled and successful in engineering. This is why our first topic in Chapter 2: Essential Skills for Success is "Be Prepared to Work Hard."

As for me, obviously, I did not quit. I (and others) thought engineering to be a great fit for my interests and aptitude, so I made the decision to change my study habits. It was not easy, and I struggled for about three semesters to re-teach myself how to study. For the longest time, I was frustrated when I entered a test and still did not know what I needed to know to perform well. I lived through some bad test grades during this time, and I had to work on not getting discouraged. None of us likes to fail, but failure is a part of engineering, so I had to learn how to deal with it constructively.

I kept working and learning, and eventually, I figured it out. Consequently, my junior and senior years were fulfilling. My study skills allowed me to become confident in the material and the application of it. I left school with a very full tool belt that has allowed me to have a fulfilling engineering career.

Mark's story highlights the fact that learning to study is one of the most important skills you must develop for success in engineering. If possible, develop good study habits BEFORE you take your first engineering course. I know you are smart and may be able to do well in courses such as Calculus and Physics without really studying, but please learn from Mark's experience. The poor study skills that enabled you to perform well in high school or in your introductory courses such as Calculus and Physics will not work with engineering classes. As described below, engineering courses are different.

- **You must take ownership of your learning.** Your engineering professors will usually be more than happy to help you in a course, but they will not be looking over your shoulder to see how you are doing. They assume no news from you means you are staying current in class and learning everything they are presenting. Successful learning in your engineering courses means accepting the responsibility for your learning. It means you have an owner's mentality for your courses and understand that learning the course material is your responsibility and not the instructor's. The instructor will not necessarily tell you everything you need to do to learn the course material. Instead, you will need to be proactive and determine what you need to do to be successful in the course. It is up to you to carefully read the syllabus and satisfy all expectations.

- **You must teach yourself.** There is too much to learn for your teacher to cover everything in class during a few hours each week. They will cover some things and then expect you to learn the rest on your own. They assume that you have learned it unless you come and ask them for help. Learning how to learn can be hard, and you will have to be accountable to yourself because there will be no one looking over your shoulder. Take ownership for learning, admit the things you don't know, and then have the discipline to sit down and learn. This is a big change from high school, but it is exactly how you will learn on the job.

- **You must integrate information.** Typically, in your math/science courses, you learn a specific technique and then use it on an exam. For example, you learn integration by parts, practice it with homework problems, and then show you can use the technique on an exam. Your engineering courses, on the other hand, teach you general principles and techniques and then require you to apply them to specific problems. Solving these engineering problems requires you to integrate what you have learned from all of your previous courses. For example, you will have to use the general principles covered in the present course you are taking, as well as geometry, trigonometry, calculus, algebra, and previous engineering courses.

- **You must work hard.** I know I have said this before, but I want to say it again. Engineering courses take an enormous amount of time, and it is likely that you are not accustomed to putting forth the amount of effort that is required to be successful in them. I routinely see high performing students have the same experience as Mark in their first true engineering course. Like Mark, some are smart and able to get by in the other courses with inadequate study habits. Engineering courses require intense discipline and force you to dig deep within yourself for the strength to persevere. Many are not prepared mentally and emotionally for the effort required. There will likely be times that you will want to quit. Your friends will all be out having fun while you are in your room studying. You will find yourself exhausted and feeling as if you have nothing left, but there are two more homework problems to complete. What will YOU do? The answer to this question has determined the fate of many engineering students, and it will determine yours. Either you quit, or you dig deep and keep on going. Remember the goal of positively affecting lives and know the hard work will be worth it. Take a quick break and clear your head, and then go back and complete your assignment. If you do this, you will realize that you are capable of achieving far more than you ever thought, and you WILL be successful.

You can search the internet and find many suggestions for how to study, but ultimately you must find what works for you. Each person is different and learns differently. Some learn well in groups, and others like to be alone. Talk to other students and learn what works for them. Research study methods on the internet. Try different study approaches and learn what works for you. The important thing is that you need to do it. Even if you do not need to study much for the classes you are currently taking, learn now. The best time to learn good study habits is when you do NOT need them, so develop good study skills in your early classes. If you do, you will have the study skills necessary when you get to your first engineering class.

> The best time to learn good study habits is when you do NOT need them, so develop good study skills in your early classes.

Get the Most from Class

For you, as an engineer, school is not simply a place to be. Your classes are where you prepare for the future by filling your tool belt. To help you fill it successfully, adopt the following guidelines.

Go to class prepared.

Read the course textbook BEFORE the material is covered in class. Yes, you heard me correctly. You should read the textbook, and you should read it BEFORE class. If you do not understand something you read, look at the examples in the book. If you still do not understand, make a note. In class, listen as the instructor covers the topic. If you still do not understand after their explanation, ask a question. Yes, you heard me correctly. Raise your hand and ask a question. I know it is hard for an engineering student to ask questions, but you must learn how. Like your manager, your instructor is there to help you when you truly need it, but you have to ask for their help. You are paying the instructor good money for the course, so allow them to do their job and help you.

Stay Engaged and Actively Listen

As an engineer, you will obtain much of the information you need to solve problems in meetings or in private conversations. A key skill you must develop is active listening. Active listening is listening with your brain engaged. You take in what you hear and then actively think about it. You process what you hear while you hear it and apply the knowledge to the current task or file it in your mental filing system for later use. If you let your mind wander and stop actively listening, you can miss a critical detail that causes your project to fail. This same concept of staying engaged and actively listening applies to your classes. To learn effectively in your engineering classes, you must engage your mind for the entire class and process what your professor says.

Staying engaged is important in school and at work, but school can teach you the bad habit of disengaging. It is easy to sit in class, let your mind wander, and think you will get the information later from the book or from somewhere else. I increasingly hear professors talk about how students sit disengaged in their classes with glazed looks on their faces. They are physically present but not mentally. The problem with disengaging and thinking you can obtain the information later is that often, later never comes, and the information is lost forever. You miss key deadlines, information for exams, and fundamental concepts necessary for further learning. You also carry this habit into the workplace, and it will negatively affect your performance.

I realize that engineering professors are not that exciting, and most of us are not very proactive at involving you in the class. This makes it easy to disengage, but you must learn how to avoid this. The next section discusses ways to stay engaged.

Take notes and practice active listening during class.

>Taking notes during class is a great way to stay engaged, and note taking will help you get more from each lecture. Many studies have shown that you retain more information from a lecture when you take notes. Even if you never look at your notes again, taking them will cause you to get more from the lecture. When you take notes, actively listen and do not write exactly what is said. Instead, stay engaged and follow the professor's train of thought. Think about and understand what is said, and then write your processed thoughts. If you are not able to understand what was said, then ask questions. Even if your instructor gives you a copy of the class notes, actively listen and write your thoughts on the notes. You are trying to master this material to have it available when called upon to use it on the job. Actively listening and taking notes helps you to make the material your own and internalize it in your mental filing system.
>
>When class ends, don't immediately disengage. Take a few minutes to review your notes while the material is fresh on your mind. This review will help you remember the information and allow you to fill in any holes or recognize where you still have questions.

I still use this habit of taking notes in meetings. Even if I never look over the notes again, I take notes in meetings to stay engaged and ensure that I am actively listening. Develop the habit of taking notes now and you will carry the practice into your career. Jonathan

Use homework for its intended use.

>Homework is not just a task to complete. Homework 1) lets you practice and see where you still do not understand a topic, and 2) lets you practice to gain the speed you will need on an exam. When you are stuck on a homework problem, the goal is not to get the problem finished. The goal is to see what concepts you do not yet understand, so you can take steps to learn the missing concepts. Let me say that again because it is very important. When you are stuck on a homework problem, the goal is not to get the problem finished. The goal is to see what concepts you do not yet understand, so you can take steps to learn the missing concepts. There are many online tools available to solve your engineering homework problems. Using one of these tools will get your homework completed, but this will not help you master the material for the course or prepare you for the next exam. Do not be short-sighted and use these tools. There are no shortcuts to mastering the concepts in your engineering courses. Mastery takes

time working directly with the concepts. Students often tell me that their grades significantly increase when they stop using online tools to complete their homework. Don't cheat yourself. Put in the work and let homework do its job of helping you to master the material. A few tips for other places to get help when you are stuck are in the next section.

> There are no shortcuts to mastering the concepts in your engineering courses. Mastery takes time working directly with the concepts.

Use Your Resources

I have often talked about how you must take ownership for learning, but being an owner does not mean you have to do it alone. Being an owner means that you take action to master the material from each course, and that often means seeking help. Below are some of the places where you can find help.

- **Peers.** Find people with your same work ethic and study together.

- **Students that are further along in school.** Upper level students have been where you are now, and most are more than willing to give you a hand.

- **Tutors.** Tutors can be a great help, but find a tutor that will understand your learning style and then help you learn. Some tutors are great at helping you complete your homework, but you learn very little from them. If your tutor is doing most of the writing, then something is wrong. Problems always make sense when someone else is talking through them. You learn engineering by personally working through problems. Even when you are working with a tutor or a study group, make sure you are not just watching others work problems.

- **Instructor.** If you are willing to do your part, your engineering instructors will be more than happy to help you. Do not go to an instructor and say, "I do not understand." You are not asking a question. You are making a statement. Typically, a statement like this means you have not put the effort in, and you want someone to tell you exactly what you need to know. You have a renter's mentality about your learning. Instead, go and say something like, "I have studied the book and the example problems. I do not understand why they perform this step. Can you help me understand why they do this?" Do you see the difference? You owned your learning. You did your part and asked for help with a specific step. Going to an instructor with a prepared question like this is similar to going to your senior engineer. Preparing the question causes you to dig deeper beforehand and may even help you solve the problem. Asking the right question is key to getting the help you need, so learning to formulate your questions will help now with office hours and will help you effectively utilize your senior engineer later. If you do your part, your instructor will be glad to help by answering a question in class or by helping you during office hours.

- **Online Learning.** There are many free online learning resources such as Kahn Academy. People learn in different ways, so sometimes, you need a topic explained differently than your instructor explains it. Since you own your learning, it makes sense to seek out other avenues when you simply cannot understand a topic. Online learning sources are an effective alternative avenue.

Next Steps

To be a prepared engineer, you must do well in your classes. Josh and Amy each used many different skills to accomplish their goal, but like all good engineers, they used one common skill – technical knowledge. You cannot be a prepared engineer without technical knowledge, so your number one priority right now is to determine how to fill your tool belt by being successful in school. Review this chapter and begin the I2-R2 Improvement Cycle. I know you are overloaded with homework, exams, and projects and it feels like you do not have time for anything else. Make time to plan ways to do well in school, or you will always feel this way. Invest the time now to develop good habits of learning, and your investment will pay great returns. Like Mark, you will become confident and enjoy learning in each course.

CHAPTER 3
REFLECTION QUESTIONS

Learning to do well in your courses is your primary concern, so use these questions to help to take an honest look at where you are and where you need to improve.

1. What do you think about the idea that the better you prepare now by learning in your engineering courses the better you will be able to perform on the job?

2. Think about your attitude toward your courses. Do you view your courses as something to complete or as a chance to master material that you will use later?

3. What do you think about the concept of treating school like a job?

4. In what ways can you see yourself in Mark's story?

5. Where are you on accepting ownership for your learning?

6. Do you know how to teach yourself, or is this something you will need to work on?

7. Have you ever done something that required you to dig deep within yourself to complete? What do you think about the idea that success in your engineering courses will require this type of effort?

8. The book talks about how it is essential to read the course textbook before you go to class. What do you think about this?

9. This chapter challenged you to raise your hand and ask questions in class. Why do you think you and other students have such a hard time doing this?

10. Why does the book say it is a good idea to take notes in class? What do you think about your current note taking habits?

11. Based on the book, what is the purpose of homework? Where are you on using homework for its intended purpose?

12. Which of the resources for getting help in a class do you need to utilize more?

13. What are two things you want to do to help you get the most out of your classes?

Chapter 4 – Begin With the End in Mind

Successful people look into the future and see their dream realized, and then they take action in the present to create the end they want. What is your dream?

In his book "The 7 Habits of Highly Effective People," business expert Stephen Covey introduces the idea of beginning with the end in mind [6]. Covey talks about how successful people look into the future and see their dream realized (the end), and then they take action in the present to create the end they want. Covey's advice is the same advice Charles Riley gave to Jesse Owens when he told him that he needed a ladder, and it is why Mark and I wrote this book.

If you are going to begin with the end in mind, then you need to understand the end. Jesse's end was clear. He wanted to be the fastest man in the world. My end was clear. I wanted to complete a sprint triathlon.

What is your end? I understand this is a hard question because you have many ends. One end is to graduate, but think beyond this. What is your dream? What do you see yourself doing eight to ten years from now? Your dream may be working in a particular industry, doing a specific type of work, or owning your own company. It is important to have a dream because it will often help you push through those tough times when you just want to quit.

Stop now and spend a few minutes thinking about your dream. Write your dream down in the box below.

This is YOUR "end." Take ownership for making it happen. Mark and I want to 1) help you understand the skills and characteristics required to make your dream a reality, and 2) motivate you to take deliberate action today through I2-R2 to build these skills and characteristics. We want to help you begin with the end in mind.

I do not know your dream, but since you are reading this book, I think I am safe in assuming it starts with you completing your engineering degree and going to work. I can say with relative certainty that your path forward will resemble the following (see Figure 5).

- You will spend the next four or five years in school preparing for your end (A).
- You will sit across the table from someone at a job interview and convince them why they should hire you over another candidate (B).
- You will be working for someone called your manager. We will talk more about them later. Even if your dream is to own your own company, you will still work for someone called investors.
- You will spend many years making a difference and continuing to learn (C).

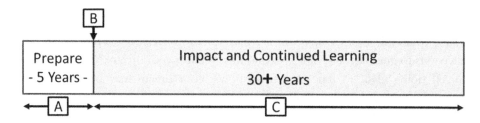

Figure 5: Broad View of Your Engineer Career

Notice the size difference between the "Prepare" and "Impact" stages. Your end will likely last for a long time — a very long time compared to the "prepare" stage. We have already talked about how the next few years in school are going to be very hard and that you are going to have to make some sacrifices. Maintain a long-term mentality during this time. Remember your dream, and make the sacrifices. You are sacrificing now to make a difference in lives. The investment you are making during school to fill your tool belt will affect your ability to perform for many years, so do not miss this opportunity to prepare well.

Look at point B. If I am correct and your dream involves working, point B is your initial "end." Your future manager needs someone to help them with their work, so they ask the interviewer to hire someone. The interviewer will be talking to many smart and qualified students just like you, and they will be evaluating each of you on the following:

1. Your technical knowledge

2. Your readiness to support your manager by accepting responsibility for and accomplishing assigned objectives

3. Your ability to get things accomplished with professionalism in a business environment

To prepare for your initial "end," you will need to develop the skills and characteristics expected by the interviewer and your first engineering manager. Most interviewers assume that your school has spent the last four years assessing your academic technical knowledge and that they would not grant your degree if your knowledge were not sufficient. Focus on doing your engineering courses well, and you will have the expected level of technical knowledge (#1 above).

> Your first end is developing the skills and characteristics expected by the interviewer and your first engineering manager.

The other two expectations (#2 and #3) deal with who you are, and this is where the interviewer will evaluate you against the other candidates. They will assume your courses have not taught you these skills, and they want to find out how you have taken ownership of developing them. If you want to stand above other candidates, then you will have to excel in these expectations.

Based on your background, you may have no idea what the interviewer and your engineering manager will expect from you. Beginning with the end in mind sounds like a great idea, but you do not know where to start. Don't worry. This chapter will help you. Do you remember how my friend Barry dealt with all of the details I had to work out for my triathlon race day? Barry knew these details would overwhelm me at the beginning, so he brought them up little by little as I was ready for them. The skills and characteristics required for success in engineering are like the details for my race day, and you will understand and work on them a little at a time. Mark and I will get you started by giving you a general idea of the skills and characteristics any engineering manager will want you to have. Generally, they will expect you to be able to accept responsibility for solving technical problems and act with professionalism. If you use I2-R2 and intentionally develop these characteristics, you will be well on your way to having what is necessary for your initial end and for making your dream a reality. An explanation of these expectations follows.

Accept Responsibility for Solving Technical Problems

Just like you, your manager works for someone, and their manager has given them specific objectives to accomplish. Your manager cannot complete all of the objectives alone, so they hired you. You were not hired to have fun or to be the president of the company. I hope your work is fun, and I believe you can be the president someday, but right out of school, your end is to help your manager by doing some of their work for them. You will advance in your career by doing a good job helping your current manager be successful. Your next manager will see your service and want you to come and help them.

How can you best serve your manager? You serve your manager by accepting full responsibility for the problems they give you and by independently taking action to see the problems through to a solution. Do you remember the discussion about being an owner on page 10? If not, stop and review it now. As an engineer, your manager expects you to be the owner of the problems they assign you. They want you to take the responsibility for a problem from them and put it on yourself.

Becoming the owner of a problem means you 1) determine the steps required to solve the problem, and then 2) execute the steps. Your manager should not have to determine the steps for you. If they determine the steps and give you small tasks to accomplish, then they are still the owner and are doing the work they hired you to do. They will gladly support you when you need assistance, but they want you to own the problem and do all you can to find solutions before you come back to them for help.

Your manager wants you to understand what they need accomplished and take responsibility for making it happen. How you accomplish the task is your responsibility. For example, consider Josh and his team. The manager did not give them a list of tasks they needed to accomplish. The manager delegated responsibility for an objective - understand what happened with the helicopter and ensure it never happens again.

> Your manager should not have to determine the steps for you. If they determine the steps and give you small tasks to accomplish, then they are still the owner and are doing the work they hired you to do.

Josh and his team determined the steps required and accomplished this objective.

To be ready to perform like Josh, understand that the interviewer will be looking for someone who can support their manager by accepting responsibility for and accomplishing assigned objectives. Begin developing this skill now by accepting responsibility for your learning.

Act with Professionalism

Your manager expects you to support them by taking responsibility for solving problems. They also expect you to act with professionalism when solving them. Do you know what it means to act with professionalism? We use the words profession and professional often, but I am not sure we always know what these terms mean. Professionals exhibit certain characteristics, and these characteristics are part of what causes us to call them a professional. Having the characteristics of a professional is the second thing the interviewer will be expecting to see in you as an engineer.

What are the characteristics of a professional that your manager and the interviewer will expect you to have? A quick literature search shows that there are many opinions on being a professional, but Figure 6 shows common characteristics always mentioned. These common characteristics (described below) represent what any engineering manager or interviewer will expect from you as an engineer, so developing them is essential for achieving your end. Many of the descriptions are going to sound familiar because we have alluded to them in some form or fashion on the previous pages. The descriptions are adapted from the MindTools Content Team article on professionalism [7]. Their website is an excellent place to start as you seek help developing specific professional areas.

1.	Specialized Knowledge
2.	Reliability
3.	Self-Motivation/ Ownership
4.	Honesty and Integrity
5.	Accountability
6.	Self-Regulation
7.	Appearance

Figure 6: Characteristics of a Professional

1. Specialized Knowledge

    Professionals are known for their specialized knowledge, and they become experts in their field. As an engineer, people are relying on your technical expertise. As in the example with Amy, they will assume you are correct and take action based on what you tell them. You must develop and maintain the specialized knowledge required to make these technically sound decisions. In school, you learn a specific topic for an exam. As a professional, you prepare for the next unknown technical challenge by learning everything you can about your field. As an engineering professional, your desire to learn should be a driving force in your daily routine.

> *Learning does not stop once school is over. Especially right out of school at your first job, having a learning mindset is vital every day. With new systems, processes, along with the technical challenges, it takes time and the willingness to learn in order to succeed.* <u>Sammie</u>

2. Reliability

    Professionals take a task or objective and complete it. They get the job done and can be counted on to do what they say they will do. If something comes up and they cannot keep a commitment, they immediately communicate and work out an acceptable solution. When problems arise, professionals do not make excuses and blame others. They take ownership and find a solution to the problem at hand.

3. Self-Motivation / Ownership

    Professionals show initiative, and they typically go above and beyond what is expected. Professionals are owners of their work, and they complete tasks without constant supervision. As an engineering professional, you will take the initiative to 1) find and solve problems, 2) learn about your job and technical area, and 3) evaluate and improve your behavior and performance.

4. Honesty and Integrity

    Professionals have an unyielding commitment to doing the right thing, and they exhibit qualities such as honesty and integrity. Professionals always keep their word and can be trusted to tell the truth. Professionals stay attuned to the ethical implications of their decisions and seek help when faced with an issue where there is no clear right or wrong answer [8]. Professionals never compromise their values. Your integrity is your greatest asset as an engineer. If your manager cannot implicitly trust what you tell them, then you are of little value to them because they must spend their time checking behind you. You earn this trust when they are confident you know what you know, and more importantly, know what you DON'T know. For what you know, they see that you take the time to think through, check, and recheck facts to ensure your answers are correct. For what you don't know, they

    > Your integrity is your greatest asset as an engineer.

    see you seeking advice from people who have the knowledge you lack. If you do not know or are not sure of an answer to a question, say, "I do not know, but I will get back to you." It is much better to say you do not know than to compromise your integrity and act as if you do.

5. Accountability

True integrity is doing the right thing when no one is looking, and professionals hold themselves accountable for always acting with integrity in their words, actions, and thoughts. When a professional makes a mistake, they own the mistake and do not blame others.

6. Self-Regulation

In his article on professionalism, Chris Joseph states the following. "A professional must maintain his poise even when facing a difficult situation. Your demeanor should exude confidence but not cockiness" [9]. The MindTools Content Team describes Self-Regulation like this. "Genuine professionals show respect for the people around them, no matter what their role or situation. A professional exhibits a high degree of emotional intelligence (EI) by considering the emotions and needs of others, and they don't let a bad day impact how they interact with colleagues or clients" [7]. Dan McCarthy adds that professionals "are polite to others and do not use derogatory or demeaning terms. A professional never bullies. There is no room for this behavior whatsoever in the workplace. Sometimes bullying is veiled in odd attempts at humor, and you know it is wrong when the humor comes at someone else's expense" [8].

7. Appearance

How others receive your technical information is often influenced by your appearance, so professionals strive to ensure their physical presentation matches the level of excellence represented in their work. First impressions can often be the longest lasting. Your appearance should always reflect the professionalism that is appropriate to your setting and those around you.

As an engineer, you are a professional, and your manager will expect you to act with professionalism. They expect you to have a basic level of professionalism when hired. They then expect you to continue to learn and develop your professional traits throughout your career.

As stated earlier, you cannot count on your courses to develop these traits in you, so you need to take action. Review this chapter and continue the I2-R2 Improvement Cycle. From the previous chapter, you should already be working on a few I2-R2 items related to being successful in school. Succeeding in school is your priority right now, so try to incorporate professional traits that will also help you perform well in school. Following are a few examples.

Self-Motivation/Ownership: Do not just do the minimum work to get by. Perform for your instructors the way you want to perform for your manager.

As a professional, you want to be known for excellent work, so do excellent work now. Your work is a reflection of you, so make sure each assignment you turn in reflects the image for which you want to be known. My father told me that it only takes a little more effort to make your work stand above the rest, and I have found this to be very true.

Reliability: If you have a problem at work, you communicate with your manager. Do the same with your instructors. If you are having trouble completing an assignment on time, communicate with your instructor. If you are going to miss work, you let someone know (trust me, you do). Do the same with your instructors. Email them and let them know if you are going to miss class. If you cannot email in advance, email them after the fact.

Honesty and Integrity: This is a big one. You will have many opportunities to compromise your integrity in school. Please do not. It is a very bad habit to start. Think about this. If you will compromise your integrity when the only consequence is a grade on an assignment, then how much will you compromise when you are at work and your job is on the line? Compromises at work often result in people getting hurt or worse, and you will face yourself in the mirror every morning knowing the compromise you made. Big compromises usually occur after several seemingly insignificant ones. Each time the compromise gets easier. Make a decision now to act with a level of integrity that avoids even the smallest compromise.

Develop these traits, and you will impress the interviewer and make your first manager very happy.

Next Steps

This chapter introduced you to the idea of beginning with the end in mind. You are on a journey to becoming an engineer, and right now, you are in the "Prepare" phase. During this phase, you are going to develop what you know and who you are. To develop what you know, focus on doing well in your engineering courses. The courses will give you your essential technical knowledge, but technical knowledge alone will not satisfy the expectations of the interviewer or your engineering manager. To satisfy their expectations, you will also need to focus on developing who you are. Learn to accept responsibility for solving technical problems and develop the personal traits of a professional.

This phase ends with you sitting in an interview convincing the interviewer why they should hire you over another candidate. What you do during your journey from now until then will determine the outcome of the interview. Your first step to making this journey a success is to make a conscious decision to take ownership of your career. If you have not done so, I encourage you to make this decision right now so your dream can become a reality.

CHAPTER 4
REFLECTION QUESTIONS

1. What does Covey's idea of beginning with the end in mind mean to you?

2. What is your dream?

3. What are your thoughts when you look at Figure 5?

4. At your first job interview, you will sit across the table from someone to convince them why they should hire you over another candidate. Which of the three things they evaluate you on needs the most work?

5. What does the book mean when it says that your manager will want you to accept ownership of the problems they assign you?

6. For each characteristic of a professional given below, think of at least one thing you want to remember about the characteristic.

    Specialized Knowledge
    Reliability
    Self-Motivation / Ownership
    Honesty and Integrity
    Accountability
    Self-Regulation
    Appearance

Chapter 5 – Practice the Discipline of WHAT Before HOW

Understanding the WHAT is like getting the compass lined up in the right direction, and then the HOW takes you there. If you jump straight to the HOW, you will likely end up in the wrong place and solve the wrong problem.

In the previous chapters, we have talked about the following foundational skills that are essential to you becoming a prepared engineer.

- Be Prepared to Work Hard
- Be an Owner
- Learn with the Intent to Act
- Be Organized
- Practice Self-Assessment
- Manage Personal Growth
- Be Successful in School
- Begin with the End in Mind
    - Accept Responsibility for Solving Technical Problems
    - Act with Professionalism

We hope you have started developing who you are by using the I2-R2 Improvement Cycle to work on developing a few of these skills deliberately. Before we end, there is one more topic to discuss, and we have intentionally saved this one for last because it is foundational to all that is good engineering. It will influence how you approach everything you do as an engineer. It will influence your success in school, at home, and at work. This skill will help you effectively develop all of the others.

Let's start with a story about a friend of mine named Bob.

Bob worked for a large technology company. The president of the company needed a special report created which required integrating data from several computer systems in real-time. Bob was assigned to the task, and he thought that this was his big chance. He met with the president to understand what he wanted and then went off and created the report. Bob was so proud of his work. He showed the report to the president and waited for the praise to flow. The president reviewed the report for a long time and said, "This is exactly what I asked you for, but this is not what I need."

Bob learned a valuable lesson that day. He learned that his job as an engineer is not just solving technical problems that give customers what they request. He learned that he must fully understand the true problem before he can solve it. He learned that the customer may ask for one thing but need something different, and they expect you as the engineer to determine and satisfy their true need. Bob learned the reality of real-world problems as illustrated in Figure 7.

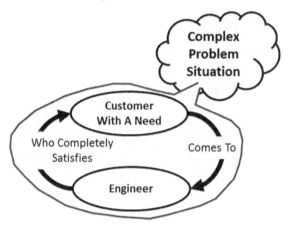

Figure 7: Real World Problem Solving

Real-world problems occur within complex situations where the customer needs something to be different. Many interacting factors create this need, and these factors tend to mask the real problem. Most of the time, the customers themselves do not fully understand their need or the real problem.

The customer does know they have a need, and they typically express this need as a problem and come to you, the engineer, to solve it.

Like Bob, this is the point where most inexperienced engineers go wrong. When faced with a problem, our natural response as engineers is to start determining HOW to solve it. We jump right to the details such as what the design will look like or what math to use to arrive at a solution. Focusing on these details is starting in the wrong place. Our first response should be to fully understand WHAT we are really trying to solve.

> An engineer does not just solve technical problems that give customers what they request. The customer may ask for one thing but need something different, and they expect you to determine and satisfy their true need.

The customer does not want us just to solve their "problem." What they really want is for us to understand WHAT their need is and then satisfy it. Stop right now and think about this expectation. Let it soak in. To the customer, we are not "Designers." We are "Needs Satisfiers." We do not just design products that meet specifications or give the customer what they request. We satisfy a customer's need, and we cannot satisfy their need (the true problem) by starting with HOW. Instead, we must start with understanding the WHAT. We should begin by asking questions such as "What are we trying to accomplish?" and "What final outcome do we want to achieve?" Understanding the WHAT is like getting our compass lined up in the right direction, and then the HOW takes us there. If we jump straight to the HOW, we will likely end up in the wrong place and solve the wrong problem. Like Bob, we may give the customer what they ask for, but we will not give them what they truly need, and their problem will still exist.

> When faced with a problem, our natural response as engineers is to start determining HOW to solve it.
>
> Our first response should be to fully understand WHAT we are really trying to solve.

After his first attempt, Bob went back and started with the WHAT. He **methodically** and **thoroughly** investigated all relevant aspects of the situation, and then **systematically** developed a solution that made the president very happy by satisfying his need. He went on to have a successful career and satisfied many customers.

To be an effective engineer like Bob, Josh, and Amy, you must learn to resist your natural tendency to jump to a solution too quickly. You have to learn to address the WHAT before addressing the HOW. I call this the discipline of WHAT before HOW. It is a discipline because it takes discipline to resist the natural pull to jump straight to HOW. We discuss this discipline here in the foundations chapter because it is that important, and it is that foundational to all you do as an engineer.

The discipline of WHAT before HOW is seemingly simple, but once mastered, it will influence every aspect of your engineering practice and is essential to your effectiveness. You will use it in school when solving class problems. It will influence how you solve problems at home and with your friends. It will make you stand out at work. One of my students, Mike, was a former Raider in the United States Marine Corps. His comments below on the Discipline of WHAT before HOW capture what I often hear from students.

> "This is incredibly simplistic and simultaneously crucial. After learning this unpretentious idea, it seemed like it should be common sense, but I can recall many instances where I have fallen into that trap throughout

the years. Now, I stop, think, and evaluate my response before considering how I will do something."

The Discipline of WHAT before HOW is described below. Study it carefully and begin now with deliberate practice to make it a normal part of your thinking process.

When faced with a problem or decision to make, the discipline of WHAT before HOW leads you to a solution using the steps illustrated in Figure 8.

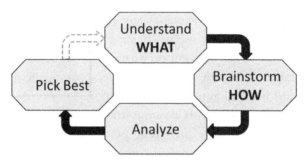

Figure 8: The Discipline of WHAT before HOW

- First, take a step back from the problem and **thoroughly** define the WHAT. Repeatedly ask "what" and "why" until you understand what it is you are trying to accomplish, and you have a clear statement of the outcome you wish to achieve. Take your time during this phase and dig deep because you limit yourself the moment you leave the WHAT phase. Assume nothing, and do not be in a hurry to move on to the "real" work. The examples below show that the "real" work for design innovation starts with thoroughly understanding the true WHAT.

- Next, **methodologically** brainstorm HOW to accomplish the WHAT. There are always many HOWs for a given WHAT. Considering these many options is what leads to effective and innovative solutions. Most design breakthroughs come while thinking creatively about the WHAT, so never settle for your first idea. Good designs emerge when many options are considered, so <u>force yourself to generate many ideas</u> before moving to the next step.

- Finally, you **systematically** analyze the various HOWs and select the one that best accomplishes your WHAT.

- When you determine HOW, it typically leads to another WHAT, and the cycle continues.

Consider the following non-engineering example that illustrates the discipline of WHAT before HOW. You are in the study room at school, and your friend comes in with a panicked look on his face and asks if you can help him. "Of course," you reply, and he tells you that he needs you to drive him across town to pick up a spare set of keys to his car. As you are walking to your car, he explains about how he went to get his fluids book out of the locked car and saw his book and keys in the front seat. He immediately panicked because he has an opened book fluids exam in an hour. He has spent the past half hour frantically trying to get into the car, and he now has decided that his only option is to go home and get the spare key. While walking, you remember that you have your book with you, so you ask your friend if he would like to borrow it. He stops and gives the familiar look of why did I not think of that. Your friend takes your book, has time to review, and then takes his exam.

Yes, this is a simple example, but it illustrates what happens to us easily in so many different contexts. Notice a few key points from this situation.

- When the problem first occurs, your friend jumps to a HOW (get into the car) and treats this as his WHAT.

    Stop and think about this situation for a minute. It is so easy to believe that the WHAT for the situation is to get into the car. His keys are locked in with his book, so he needs to get in the car. Right? No, your friend's immediate need is to have a book to use during his exam. Having a book for his exam is his WHAT. Getting into the car is one HOW.

- His focus on the wrong WHAT of getting into the car is distracting him from his true NEED (having a book for the exam).

    Getting in the car is just one way to obtain a book for his exam, but notice that he is not considering any other ways. Improperly defining his WHAT has given him tunnel vision, so he is not thinking about anything other than one HOW - getting into the car. With this mindset, his activity seems to make sense to him. In reality, he does not have to get into the car. He does need a book for his exam, and trying to get into the car is keeping him from doing this. Not understanding the WHAT always distracts you from solving the true problem. This is why it is so important to take your time during the WHAT phase.

- Clarifying the true WHAT changes everything.

    It changes everything when you look at the situation in light of the true WHAT of obtaining a book for the exam. All of a sudden, spending time getting into the car does not seem like such a good idea. With a clear picture of the true WHAT (obtaining a book for the exam), it is obvious that there are many other ways (HOWs) to achieve the WHAT. He could see if the instructor has a book. He could see if he can use the online version. He could see if a friend in another section has one. He could see if someone can bring him a spare key. When the focus moves from a HOW to the true WHAT, the tunnel vision opens up, and the solution space instantly goes from one option to many options. A clear focus of the WHAT always does this.

- Your friend (the customer) needs your help to understand his true need.
  Like the customers you will work with, your friend is in the middle of a complex problem situation, and he needs you to help him understand his true need. The problems you will face as an engineer will be much more complicated and will have many more dimensions than this situation, but the customer needs you to bring clarity just as you did for your friend in this simple example.

You may look at this example and think that you would never be as blind as the friend was, but engineers do it every day. It is easy to miss the WHAT and jump right to a HOW, but doing so can easily end in disaster. Remember Bob? Bob skipped understanding his WHAT and focused only on HOW. Skipping the WHAT caused him to develop a very good solution for the wrong problem, and the customer's need was left unsatisfied. This is what typically happens when you skip defining the WHAT.

I'm so glad that this topic is mentioned in your book, because this is something that most engineering students are never truly made aware of until they experience it in real life. As Mike mentioned, it's something that even experienced professionals don't always recognize until explained to them. It is a crucial skill and mindset for students. Alina

The examples below illustrate the discipline of WHAT before HOW and show how important it is to effective problem solving.

Design Innovation in the Classroom - Robot Runner

The first example to help you better understand the discipline of WHAT before HOW is from a design competition in my freshman engineering class called the Robot Runner. For the Robot Runner competition, students must design and build a vehicle to maneuver from point A to point B in a maze as rapidly as possible. The exact dimensions are not given, but the maze will have a similar shape to the one shown in Figure 9.

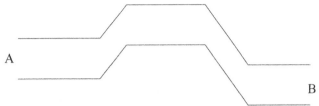

Figure 9: Robot Runner Example

Most students define the WHAT as "build a robot to maneuver the maze" and create a smart device that senses the sides of the maze, turns, and makes its way through the maze. By skipping the WHAT phase, the students pass over all possible solutions for quickly getting from point A to point B except one – a smart device. No matter how well they design, all they will ever develop is a robot. With no evaluation, they eliminated all possibilities and narrowed their focus to one alternative. This narrowing of options is what skipping the WHAT always does. When you jump straight to HOW, you bypass the analysis of options that leads to the best solution, and doing so too early can be very costly.

One group applied the discipline of WHAT before HOW and defined their WHAT as "build a vehicle which gets from point A to point B as quickly as possible." They brainstormed many options and came up with the design shown in Figure 10.

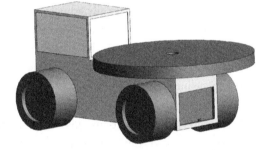

The vehicle was equipped with a strong motor and a disk on the front that rotates. When the disk contacts the side of the maze, the strong motor causes the disk to rotate and follow the wall. The vehicle forced its

Figure 10: WHAT Design Robot Runner

way through the maze in a matter of seconds. In fact, it was finished before all other competitors even sensed the first turn and adjusted the course of their vehicle.

The design blew everyone away. Notice where the team won. The team "beat" the others during the WHAT phase. All of the other teams defined their WHAT as "build a robot to maneuver the maze," and that is exactly what they created. They completely ignored the first three steps in Figure 8 and focused on one HOW – a robot. They cheated themselves by blindly ignoring all possible solutions except one. This is what happens when you jump straight to HOW.

In contrast to the other teams, the winning team applied the discipline of WHAT before HOW. They took a step back and first focused on defining WHAT they were trying to accomplish - "get from point A to point B as quickly as possible." Instead of locking into one narrow idea, they followed the steps in Figure 8 and considered many possibilities for achieving the WHAT. An innovative solution emerged. You will see this pattern of innovative design breakthroughs coming from a focus on the WHAT in the next example.

> They cheated themselves by blindly ignoring all possible solutions except one. This is what happens when you jump straight to HOW.

Design Innovation in the Workplace

In this example, I talk about two companies that changed the world by using the discipline of WHAT before HOW. They both were in markets with clearly defined products. For years, their designers had focused on making their product better at doing its job. For whatever reason, both companies did the same thing. They stopped and tried to understand the WHAT. Instead of asking, "HOW can we make our product better at doing its job?" they asked, "WHAT is the true job of our product?" Changing the question changed the world, and you are benefiting from the results today. A simplified discussion of each scenario follows.

The first product is the point and click camera. You may ask, "What is a point and click camera?" A point and click camera is like the camera on your phone. You point it, click it, and it automatically adjusts light and focus to ensure you obtain a quality picture. Before point and click technology, taking pictures was a complicated task. First, you would measure the light level and adjust the lens to allow the proper amount of light to contact the film. Next, you would make adjustments on the lens to bring your image into focus. Every camera worked this way, and designers worked to create better cameras. They made improvements to things such as lenses, finer light adjustments, and better focus aids. In graduate school, I heard the story of how one company changed the question and asked, "What business are we in?" Most responded with the HOW of making cameras, but as they dove deeper, they realized that their true WHAT was to give people the ability to capture memories. This changed everything. They went into people's homes to see how they were doing at allowing them to capture memories, and they found pictures out of focus, with improper light level, and off center. They now knew that to accomplish their WHAT of providing people the ability to capture memories, they needed a camera that would give a quality memory by just pointing and clicking. Clarifying the WHAT gave birth to the vision for the modern camera that we all know. Through the years, I have captured many good memories with my point and click cameras, so I am glad they practiced the discipline of WHAT before HOW. Aren't you?

The second product is a little more familiar. It is the smartphone. This may come as a surprise, but Apple was not the first to create a smartphone. Before Apple, Blackberry dominated the market with a phone that provided mail and messaging. The problem is that they thought their WHAT was providing a phone. They focused on making a better phone with improved screens, keyboards, and services. Apple changed the question. Instead of seeking to design a better phone, they defined their WHAT as design a personal entertainment center. We all know the result.

Manufacturing Plant Automation Project

This example of the discipline of WHAT before HOW is from my first job out of graduate school. A large cabinet company hired me to automate one of their wood processing facilities, and my manager made it very clear that my mission was to automate the facility successfully. I could have jumped in and started working on all of the details for HOW to automate the various processes. Instead, I began by understanding the WHAT. I learned that the plant received orders for parts, such as doors and drawer fronts, from several cabinet assembly plants. Rough lumber was purchased from lumber mills, and the plant processed it to produce the cabinet parts shipped to the assembly plants.

As I explored the WHAT, I learned that the plant was facing critical changes in both of its key inputs (lumber and orders). Lumber prices were rapidly increasing, and the quality of the incoming lumber was simultaneously decreasing. At the same time, the assembly plants were ordering more frequently and in smaller batch sizes.

It became clear that the true WHAT was not the initial request to automate the facility. The true WHAT was to maintain the cost competitiveness of the plant despite the changes with lumber and orders. With a clear understanding of the WHAT, we designed and implemented an automated facility that did satisfy the true need. The facility provided the flexibility necessary to accommodate orders from the assembly plants, and it maintained cost competitiveness despite the rising lumber prices and decreasing quality.

If I had only done what I was asked, "automate the plant," I would not have done it in a way that would have adequately addressed the lumber and order issues. I would have given the customer (my manager) what he asked for, but I certainly would not have satisfied his NEED, and the project would have been a failure. Using the discipline of WHAT before HOW saved the project and my job.

Josh and the Hard Landing of the CH53E

For the final example of the discipline of WHAT before HOW, recall the story from Chapter 1 of the CH 53E that had to make a hard landing. Josh's team quickly determined that a condition that developed over time forced the aircraft to make the hard landing. At this point, it would have been easy for them to turn their attention and get lost in the technical details of determining how to eliminate the condition. Instead, the team stopped and remembered the WHAT – ensure that everyone involved with this helicopter remains safe. When they practiced the discipline of WHAT before HOW, they realized the first course of action was to implement a reoccurring inspection of all aircraft to identify the condition and fix it before it became a problem. This action accomplished the WHAT of ensuring everyone remains safe. With an initial solution to their WHAT in place, they were free to move to the next WHAT - find a permanent solution by exploring the technical details of how to eliminate the condition. Using the discipline of WHAT before HOW kept the team focused on the key issue and kept the pilots and crew safe.

If you want to be a good engineer, start now developing the discipline of WHAT before HOW and let it drive all that you do. Below are a few examples of what this may look like in the context of school.

> It is time to sign up for an engineering course, and you need to decide what section you want to attend. You begin by going to your school's website to review the historical grade distributions for the various professors teaching the course. While reviewing the grades, you remember the discipline of WHAT before HOW. You stop and ask yourself what you are trying to accomplish. At first, the answer seems obvious. You are trying to avoid the professors with the lowest grade distributions. As you probe deeper, you realize that the true WHAT is finding the professor that will best fill your tool belt. You do not simply want a good grade. You want to learn the course material so that you can use it. Reviewing grade distributions is only one HOW, so you think broader and seek additional information from people who have taken the course to determine the professor who will help you learn the most.

> You are stuck on a homework problem, and the homework is due in a few days. Your professor does not prohibit the use of online resources for solving homework problems, so you submit the problem and receive a solution. You review the online solution, complete your problem, and think you are done. Then you remember that the WHAT is not to complete a homework assignment. The WHAT is to allow the homework to help you master the course material. With an understanding of the WHAT, you realize two things. First, you see that submitting the homework for an online solution is only one HOW. You can also go to an instructor's office hours or get help from a classmate or someone that has already taken the course. Second, with a

clear understanding of the WHAT, you go back to the online solution to understand why you could not solve the problem. You look for the concepts you were missing and then make sure you understand them now. You let the online solution help you master the course material.

As you can see from these examples, practicing the discipline of WHAT before HOW requires you to slow down and look at the big picture. I understand that this is not easy. Learning this discipline will take some deliberate practice, but you need to put forth the effort now and learn it. Mastering the discipline of WHAT before HOW is likely the most significant thing you can do to prepare for your career as an engineer.

CHAPTER 5
REFLECTION QUESTIONS

1. What comes to mind when you read the story about Bob?

2. What do you think about the idea that you are a "Needs Satisfier?" How do you think knowing this can change how you approach problem solving?

3. How would you explain the discipline of WHAT before HOW to a friend?

4. Which example illustrating the discipline of WHAT before HOW did you like the most? Why?

5. What can you do to start practicing the discipline of WHAT before HOW?

Chapter 6 – Final Thoughts

"The only impossible journey is the one you never begin" – Tony Robbins

---

This is Mark, and I congratulate you. Reading this book means you have begun your journey to establish your foundation for a successful career in engineering. This is a great start, and you should celebrate your success. In fact, Bill and I believe that if you have diligently worked through the material in this book, you now have an appreciable head start on long-term classroom and career satisfaction and success.

Like any trip or journey through a land that you have not traveled before, it helps to have a guidebook that gives you a starting point and sends you on your way. Use this book as such a guide. Keep it close. Dog-ear the pages to which you find yourself continually returning. We've provided some of OUR key takeaways in bold print and highlighted boxes throughout the text, but feel free to highlight your own – the ones that speak to your personal development opportunities. As you progress through school, review this guide each semester and allow it to keep you on the right path. When you feel the material in this book becoming a part of who you are, use the I2-R2 Improvement Cycle with the material in appendix A and appendix B to continue developing the traits of successful engineers.

Before we go, I want to share a few thoughts.

You have undoubtedly heard the term "soft skills" before. If you are like me, the word "soft" denotes optional or nice-to-have. That is not the right connotation for the skills Bill and I have shared with you. Instead, we hope that from now on, you view the skills shared in this book as not "soft skills" but rather as "success skills." You should view them as essential instead of optional. I have worked with many engineers, and it is my observation and experience that an academically solid student that has a mastery of many of these success skills will have a more fulfilling and successful career than the

> An academically solid student that has a mastery of many of these success skills will have a more fulfilling and successful career than the academic super-star who either chooses not to develop success skills or isn't aware of them.

academic super-star who either chooses not to develop success skills or isn't aware of them.

As you apply what you have read, note that the skills are not "add-ons" to your daily activities. They are not something extra you "do." Rather, they are a change to your habits, and ultimately a change in the way you behave and how you perform your daily duties – in the classroom and later in the workplace. By working on these skills, you are developing who you are. We are not suggesting you be someone you are not, but we are suggesting that you become a better and more complete version of who you already are. It is our belief and experience that this level of development will provide you lasting improvement.

As your journey continues, we remind you to:

Using the guidance provided in this book, **methodically** plan a course of action for your educational journey and your career that will follow.

Thoroughly develop that plan to enable YOU to succeed. Utilize the I2-R2 Improvement Cycle. Be specific.

Systematically work through that plan. Make it a part of your routine. Lean on a success partner along the way to hold yourself accountable.

Chapter 3 introduced you to the idea of beginning with the end in mind, and you wrote down your end (your dream). Internalizing the information in this book will help you achieve your dream and will prepare you for your initial end of securing your first job. Take responsibility for your success and climb the ladder one step at a time. Climbing will be hard and there will be sacrifices to make over the next few years, but keep a long-term mentality. Your journey will come with a series of vector changes, so be willing to accept small but measurable changes early knowing it will result in greater progress as you move forward. Remember how Bill described preparing for his triathlon? The training was tough, but it caused his race day to be fun and very fulfilling. Put the work in now to prepare, and you will find the same joy and fulfillment as you see your dream turn to reality.

Bill and I are humbled that you have taken the time to use this book. We thank you for entrusting us with your development, and we hope that you will reflect upon what you have learned and utilize it as a ready reference for years to come. We are confident in you.

Be well and let the journey to becoming a prepared and successful engineer begin! You can do it!

Help Shape the Next Volume of
Preparing for a Career in Engineering

Your journey to becoming a prepared engineer is just beginning, and Mark and I have more we want to share with you as you progress through school. We are working on volume two of *Preparing for a Career in Engineering,* but we need your help.

Standing on the other side of your journey allows us to see and understand things that you cannot. We will use this advantage to share insights with you in volume two, but our view is limited, and we need your insights as well. Standing on the other side gives us perspective, but it also blinds us. We are not in your position, and it is easy to forget what it is like. We know what we want to tell you, but we do not know what you want to learn. Without your help, volume two will be incomplete.

To help us understand you and your needs, please use the QR Code or link below and share your insights with us.

https://forms.gle/pGBJvoqjoRuxzfZ39

Appendix A
Differences Between School and Work

Mark and I interviewed many engineers and asked them the key differences between school and work. Their responses are in the table below. Use this table to understand the mindset you will need on the job and begin developing it now. Notice that many of the work items require confidence to achieve. Perform well in your engineering courses and build this confidence in your knowledge and in yourself.

SCHOOL –VS- WORK

SCHOOL	WORK
Learn only when there is an assignment to complete	Take the initiative to learn to expand knowledge and capabilities for the next unknown challenge
Follow rubric and do what I am told	Do what is needed to accomplish an objective – requires confidence
Solve the problem and deliver an answer	Accomplish a given objective
Use this technique to solve this problem as I showed you	Adapt the technique to most effectively accomplish the objective. Do what you think is best – requires confidence
What I am **supposed** to do on this assignment	You determine what is needed to accomplish the given objective and then do it – requires confidence
Always do this (use 3 decimal places, use this algorithm for sorting,)	Use engineering judgment to determine what does and does not matter in accomplishing the given objectives – requires confidence
Always told how you performed on each task – grade	Feedback comes from self-assessment. External feedback typically comes only after there is a problem and usually after it has persisted for some time
Disengage in class and pick it up later	You must stay completely engaged during conversations and in meetings, or you will miss information that cannot be obtained later
Someone will tell me what I need to do and when I need to act	You are the "someone" – requires confidence

Used to being at the top of the class and knocking everything out of the park – or at least performing well	Things are very different now. You are not an expert in this new environment. There is way too much new and way too many factors influencing every problem for you to be an expert. So, 1) be a learner. 2) Cut yourself some slack. Failure is normal and is how you grow
Work to get a grade and receive instant affirmation	Work and achieve an objective which may or may not be noticed by those around you
Solve problems with other smart people around to lean on and check your work	Solve problems where you may be the only engineer involved, and you must provide technical guidance – requires confidence
Your main focus is accomplishing your work	Your main focus is serving your manager by accomplishing some of their work
The instructor is responsible for teaching you what you need to know	You are responsible. You take the initiative to 1) accomplish assigned objectives, 2) learn about your job and technical area, and 3) evaluate and improve your behavior and performance
You submit work expecting feedback and a chance for revisions	The quality and completeness of your work should be such that you expect your manager to approve it and send it on
You are expected to have an answer to every question asked	You are expected to say, "I do not know, but will get back with you" when you are not absolutely certain
You figure everything out yourself and rarely go to the instructor	You have to know when to get help. You earn trust when people are confident you know what you know, and more importantly, know what you DON'T know. For what you know, they see that you take the time to think through, check, and recheck facts to ensure your answers are correct. For what you don't know, they see you seeking advice from people who have the knowledge you lack
Finite Problems and Solutions	Real-World problems without neat solutions.
Wrong answer can earn a 90% and result in an "A"	Wrong answers and sloppy work cost money and result in negative impact on lives
The recipient of your work and communications is your instructor	The recipient of your work and communications varies. You must account for the differences of each audience and adjust
Semester mentality with a break at the end	Must achieve life/work balance because you are in it for the long haul
Homework and exams are often centered around finding the correct answer	There may be many solutions that address an issue, and critical thinking must be used to determine the best solution given the entire circumstance

Appendix B
Success Traits of an Engineer

To understand the common traits of successful engineers, Mark and I worked with several business professionals responsible for hiring and managing engineers. Mr. Ramsey Davis summarized these traits, and an expanded version of his list follows. Once you are on your way implementing the foundational traits in this book, use these descriptions to understand further the traits of a successful engineer and continue the I2-R2 Improvement Cycle.

1. PROFESSIONALISM

   a. *Promptness (i.e., meetings and deadlines).* It is important to keep in mind that everyone around you has priorities and responsibilities, and when you are late, you disrespect them. Your actions play a part in helping or hindering those around you. When deadlines are not met, or meetings start late, you impact others, and it can have a cascading effect. If a meeting is important enough to attend, many times some of the most important information will be discussed at the opening. You should plan to be 5 – 10 minutes early. The time before meetings can also be used to introduce yourself and speak to those you do not know, and it is an opportunity to make a good first impression.

   b. *Personal Appearance.* Personal appearance can affect how your technical information is delivered and perceived by others. First impressions can often be the longest lasting. Your personal appearance should always reflect the professionalism that is appropriate to and reflects the importance of your settings and those around you.

   Be aware that you never know who you may run into when you are on campus. You can easily meet someone who will be interviewing you for a job in the future. Always keep this in mind. You should dress for the job you want rather than for the job you currently have. If you want to be an engineer, then start dressing like one.

   c. *Communication Etiquette.* Professional communication is the key to showing respect to those around you. Information is most successfully delivered when done appropriately and professionally. Workplace communications many times will have a basic format and standard operating procedure, especially email and telephones, so research and learn it quickly and early.

For example, if you are in the "To" line of an email (versus being "cc'd"), even if no information is requested, it is common courtesy to reply to an email to acknowledge the email was received. When responding to emails, be sure to use complete sentences, proper grammar and punctuation, and no "texting" shortcuts or abbreviations. Email is official correspondence so treat it just as you would the body of a letter.

2. TECHNICAL CREDIBILITY

   a. *Makes and defends sound technical decisions.* Technical discussions will happen often and regularly. Sometimes these discussions will happen with no warning. Your ability to listen, decide, deliver, and defend will have the greatest effect on the technical directions of any given project. Your decisions need to be technically sound so that you can defend them when others challenge them. Always be prepared to defend the decisions, but also be willing to admit if the initial decision was not the best. Your peers are not challenging you when they pick your ideas apart. They are doing what you want them to – ensuring you have not missed anything and that your implemented solution will be successful

   b. *Seeks help when it is needed.* Seeking help to gain technical knowledge and judgment when there is no right or wrong answer is an important part of being an engineer. Homework and exams are often centered around finding the correct answer, while in a work setting, there may be many solutions that address an issue. Critical thinking must be used to determine the best solution given the entire circumstance.

   c. *Understands people will assume you are right and ensures answers are correct.* Others are relying on your technical expertise as the basis for their decisions. They will take action based on what you tell them. Take the time to think through, check, and recheck facts to ensure your answers to technical questions are correct. If you are not sure, then make sure you communicate this. If you do not know or are not sure of an answer to a question, say, "I do not know, but I will get back to you." It is much better to say you do not know than to give an answer that is wrong and then have someone act on it.

      Be prepared to answer questions about your work. Some questions are simply to give the listener (who does not have your expertise) an understanding of what you are presenting. Other questions (from those who do have expertise) are a process of becoming comfortable with and validating your analysis/design. Do not be defensive or put off when others question your work. If you have done your job, you are the expert in what you are presenting and will be able to answer the questions.

    d. *Admits when a flaw is found in an idea.* You will often present your ideas to your peers, and they will challenge them (just like you, they think their idea is best). There is nothing wrong with defending your idea, but keep an open mind and truly listen to what they have to say. You have to be willing to let go of your idea and say, "You are right." If someone finds a problem that you do not have an answer for, admit it and move on to plan B. Do not keep arguing just to "win." The goal is always to find the best solution to the problem, regardless of whose idea it is.

3. WORK ETHIC

    a. *Goes above and beyond what is expected.* Many times a project can be completed simply and quickly; however, it is important to be known for a work ethic that ensures the project has been completed in its entirety. It is not enough to do the minimum to get by and simply answer the question at hand.

    b. *Is eager to work and learn.* Once a person has found a subject that interests him/her, it should be obvious to those around them. Your willingness to go the extra mile to learn a subject matter that is in your area of interest should be a driving force in your daily approach to achieve perfection. Also, know that learning never stops. Your diploma only certifies you to continue your learning in the workplace. No one has all the answers, and you will not be the first.

    c. *Takes advantage of opportunities to improve or get involved.* It is important not simply to be involved but to seek out opportunities to be involved. A person is much more advantageous to the group when he/she seeks out things to do.

    d. *Takes the initiative and works independently.* It is important to know how you fit into the working group. As you progress through projects, people are always around to help and answer questions. However, questions should be well thought out and properly researched before seeking help/clarification. Knowing how much individual effort to put in before reaching out for help is always a moving target, but your ability to grasp this concept will make you a more valuable teammate. Everyone around you will have their own tasks, so it is important that you are doing your part independently as much as possible. Know your boundaries, but be confident enough to ask for help when you need it.

e. *Effectively utilizes available resources.* Every project will have various tools to help the group complete the task at hand. Knowing what tools are available and how to utilize them best will lead to more success. A successful person will actively seek out better and quicker ways to achieve professional answers to the task. This can also play a role in how information is presented to those around you. If a tool is available and can communicate information more accurately, the most successful person will use these tools even if it takes a little extra time and effort.

f. *Takes a task/objective and completes it.* Those that hand you a task are depending on you to complete it quickly and to the best of your ability. Many times your task will be part of a bigger project, so a single failure on any task can make the entire project a failure. You will always be part of a team, and your part in any project is important, even if it appears on the surface to be negligible. Project leads are depending on all their team members to deliver quality products so that the project is delivered on time and to the best quality possible. Refer to the discussion above on "accept full responsibility for a problem and take action to see it through to a solution."

4. ABILITY TO COMMUNICATE

   a. *Can communicate technical ideas and information in writing.* A person's ideas and information are only as valuable as how well they can communicate them. The ability to write clearly and directly is the key to communicating information. An engineer will not only be responsible for design processes, but they will then have to communicate the final design to the people who require the information.

   b. *Can communicate technical ideas and information through a formal oral presentation.* Many times technical information must be delivered as a presentation to people who have very little background on the information that is being presented. A person's ability to explain the relative importance of the information and how the final product was derived will many times set the level of success of the project.

   c. *Can communicate technical ideas and information through informal communications.* In an engineer's daily life, informal technical communications or conversations will happen regularly. It is often those times that the most pertinent ideas originate. A person's ability to quickly understand and participate in these discussions will not only benefit them but often those around them as well. You must be comfortable communicating with peers as well as people above and below you in the organizational chart.

You will find yourself working with people who are lower than you on the organizational chart. Their position on the organizational chart is irrelevant to their ability to contribute. People who work the closest to the problem (artisans, shop floor workers) have insights into the problem that you will never have. They are a wealth of knowledge, and they deserve your utmost respect. Give them this respect, and they will gladly help you. You also may find yourself somewhat intimated when having to talk with them. This is normal. You are an engineer, and you like to feel competent, but you may not feel this way in the production environment. Do not compensate for your insecurity with a false sense of confidence that will come across as arrogance. Instead, respectfully ask questions, actively listen, and learn. Pairing your technical knowledge with the intimate practical knowledge from these people is how you create great solutions.

5. ABILITY TO WORK WITH PEOPLE

   a. *Understands different personality types and appreciates/leverages those differences.* A school atmosphere is a great opportunity to hone your skills at working with different personality traits. In the workplace, you will work with people of differing personalities and differing strengths. This is often by design. These strengths can be used together as a team to deliver higher quality products to customers. Your ability to recognize and take advantage of different personalities in a team will make you and those around you more successful.

   b. *Can work in a group with people even if you do not like everyone.* Maintaining a productive working atmosphere is important. Your ability to work with others, even if personality differences make that difficult, will be important for your success. Work environments are much more constrained than school, so it is important to learn how to work well with people. It is great when everyone gets along, but work is not about being social and hanging out. It is about delivering a quality product.

   c. *Can effectively deal with conflict.* Personality differences, technical differences, and opinion differences are all common sources of conflict. Conflict can often arise from many different interactions. How to work through conflicts efficiently and effectively will be a skill that can lead you and your team to great success. People will come to you for help if you can show that you are capable of working through conflicts, which many times can help propel projects to greatness. Some tips for understanding and dealing with conflict are given below.

<u>Learn not to take it personally when someone challenges your idea</u>

When someone challenges our idea, we can respond in one of two ways.

> **Response 1:** We can realize that the challenge is a normal part of the engineering process and have an adult conversation with the person to arrive at the best solution. We will hear their criticism and seek to offer a logical defense when appropriate. We will have a healthy dialogue. When they uncover a true flaw in our idea, we will acknowledge it and gain insight by talking through the issues with them. In the end, either we will abandon the idea, or we will take their insights and improve our idea by modifying it.

> **Response 2:** We can take the challenge to our idea personally and then feel the need to defend ourselves and show that we are right. In doing this, it is likely that we will get angry. Our anger will come out in one of two ways. Our anger may cause us to dig in even harder with our attempt to show how we are right and that the other person is wrong. When this happens, all logic is out the window, and the problem at hand is secondary to winning. Emotions are likely to escalate. The other way our anger can come out is by withdrawing. We get our feelings hurt and check out. We take our toys and go home. The result is the same whichever way we respond. We hurt the effectiveness of our team, and we are no closer to finding the best solution for the customer than when we started. Both you and the customer lose in this situation.

<u>Always remember that the goal is to find the best idea. It is not to win!</u>

- As an engineer, you like to be right. We all do, and that is okay, but you have to realize that you are not going to be right all the time. You are in school with many other smart people, and they will be right sometimes as well.
- It is never about winning. It is always about finding the best solution for the customer.

<u>Actively listen and fully understand the other person's point.</u>

- Actively listen to the other person to understand their point. The other person is smart, and you want to make sure you understand their insight.
- Do not formulate your next point while the person is still talking. Formulating your point while someone is talking is a "winning" mentality, and you should avoid doing it.
- When the other person finishes talking, don't be afraid of the silence that may occur as you think about what they said.
- To make sure you understand the other person's point, ask clarifying questions or summarize what you think they are saying.
- Acknowledge when the person has a valid criticism of your idea. Hear the good points they make and incorporate their insights into your idea.

<u>Stick to the facts</u>

- Engineers make decisions on facts, not on opinions.
- Separate the facts from opinions. It is easy to say, "I do not think that will work," but that is an opinion. When someone gives an opinion, ask questions to determine the facts. For example, "What about it do you think will not work?" or "Why will it not work?"

<u>Make comments in a constructive way</u>

- Think before you speak. I think we all know what this means.
- When possible, phrase your feedback in a positive tone. Below are a few examples.
    - Say something like, "Could you explain how that would work?" instead of "That does not make any sense to me."
    - Say something like "I must have missed when you told us the pipe diameter, can you please give it to me" instead of "You forgot to give us the pipe diameter. What is it?"
- Criticize ideas and not the person. Realize that the other person is not criticizing you. They are criticizing your idea, and you need to separate the two.
- Don't just find problems. It is easy always to be negative and find problems with ideas. Along with identifying problems, try and find a solution to your objection.

<u>Make comments in a way that promotes discussion</u>

As engineers, we can often be very direct and state our opinions as a fact. When we make opinion statements as facts, we shut down communications and can easily put the other person in defensive mode. Imagine you are in a group, and you have just shared your idea for solving a problem. One of the group members says, "That will not work." The team member has created a situation of you versus them. They did not open up a discussion on your idea where the unique perspectives from the group could contribute. They created a situation where one of you is right, and one is wrong. This environment is not healthy for your group. Instead of saying, "This will not work," what if they said, "Can you explain how that idea will be able to overcome this issue?" Notice the difference? The second way opens up communication instead of shutting it down. Try to avoid closed-ended statements that leave the other members nothing to say. Instead, learn to ask good questions.

<u>Know when to take a break</u>

At times, group discussions can get a little intense, and you may feel your emotions rising. When this happens, the best thing to do is to take a break. Tell your group that the discussion is getting a little intense for you and that you are going to take a break. Walk away, calm down, and then come back ready to contribute. Walking away is a lot better than having to apologize for saying something that you really did not mean.

d. *Can lead a group of peers.* Leadership can come in many different forms. At different points in a project, the leadership role can shift from one person to another depending on the task that needs to be completed. Leadership is a trait that is difficult to perfect or learn, but it is often the result of mastering some of the earlier defined traits. Proper leadership can greatly influence the success of your career and your projects. Remember, true leadership is serving the group to accomplish the task.

e. *Can follow a peer leader.* Just as it is important to be able to lead, it is equally important to be able to follow. You must have the ability to recognize when you are the leader and when you are the follower. Your ability to take action on a leader's direction can often make a big difference in the success of a task. However, blindly following direction is NOT what is needed either. This often leads to failed projects and can reflect poorly on your technical ability. Your input and technical advice should always be a driving force in your direction. If a leader has tasked you with something, and you think the direction is wrong, then part of your task is to take a direction change back to the leader.

f. *Understands how to respect the rank/position of superiors appropriately.* It is important to know the rank/position and to show proper respect of those around you. This includes others that may not be directly in your supervisory "chain," i.e., civilian engineers working with military customers. You will interact with senior personnel many times during your career. Knowing your place in the "food chain," coupled with a strong work ethic, will go a long way to earning respect from your superiors.

Your ability to quickly recognize senior leadership and adapt to the situation is also important to your success. People have worked hard to get to the upper level of their careers, and this should be reflected in how you interact with them. For example, if you are a student, you should always address an instructor with a Ph.D. as "Doctor."

Being a superior doesn't make them always right, but it does mean they are in charge. There are ways to communicate differences in opinion or direction and remain respectful to your superiors. Be respectful of their opinions, but be confident in your ability to present a different opinion respectfully. Points of view that are well thought out and presented with skill are often well received. This is a skill that will lead you to great success in your career.

Works Cited

[1] I. M. E. F. Courtesy Story, "MARINES: THE OFFICIAL WEBSITE OF THE UNITES STATES MARINE COPRS," 4 September 2015. [Online]. Available: https://www.marines.mil/News/News-Display/Article/616282/marine-killed-in-ch-53e-mishap-identified/. [Accessed 26 December 2019].

[2] J. Owens, "1936: Golden Moment of Triumph," 19 February 2016. [Online]. Available: https://www.saturdayeveningpost.com/2016/02/1936-golden-moment-of-triumph/.

[3] C. Geoffre, "What it takes to be great," 19 October 2006. [Online]. Available: http://archive.fortune.com/magazines/fortune/fortune_archive/2006/10/30/8391794/index.htm.

[4] R. Glazer, "Friday Forward," RobertGlazer.com, [Online]. Available: https://www.fridayfwd.com/blame-game/.

[5] D. J. Brown, The Boys in the Boat, New Youk, NY: Penguin Publishing Group, 2013.

[6] S. Covey, The 7 Habits of Highly Effective People, New York, NY: Simon & Schuster, 1989,2004.

[7] MindTools Content Team, "Professionalism: Developing this Vital Characteristic," MindTools, [Online]. Available: https://www.mindtools.com/pages/article/professionalism.htm.

[8] D. MCcarthy, "Understanding the Behaviors that Showcase Workplace Professionalism," thebalancecareers, [Online]. Available: https://www.thebalancecareers.com/what-is-workplace-professionalism-2275961.

[9] C. Joseph, "10 Characteristis of Professionalism," [Online]. Available: https://smallbusiness.chron.com/10-characteristics-professionalism-708.html.

About the Authors

Dr. Fortney was trained in advanced manufacturing at Purdue University and received his Ph.D. in Industrial Engineering from the University of Tennessee. At UT, Fortney's research focused on systems approaches to organizational improvement, and after graduation, he utilized the systems perspective to design and install automated manufacturing systems for MasterBrand Cabinets. Fortney's work at MasterBrand's Crossville, TN plant resulted in a system called "the rough mill of the future" and was featured in the November 1996 issue of Wood & Wood Products. His other roles at MasterBrand included corporate industrial engineering manager and general manager of the Kinston, NC Facility. As general manager, Fortney used the systems approach to design, create, and operate all aspects (human resources, production control, manufacturing) of a new 600,000 square foot cabinet facility.

Since MasterBrand, Fortney has been with North Carolina State University developing a new site-based engineering program focused on the unique needs of the Fleet Readiness Center East Research and Engineering Group for the Naval Air Systems Command. His ABET Accredited Mechanical Engineering Systems (MES) BSE program now provides NAVAIR with a steady stream of highly qualified mechanical engineers for their Havelock, NC facility. While creating the MES program, Fortney has focused on developing techniques to teach the systems perspective for design to undergraduate mechanical engineering students. He has embedded these techniques within his MES senior capstone design course as well as within a junior level project-based course where students learn and experience the System Engineering approach.

Outside of work, Fortney enjoys serving at church with his wife and playing with his many grandchildren.

Mr. Mark D. Meno is the Maintenance, Repair, and Overhaul (MRO) Engineering Director at the Dept of Navy's Fleet Readiness Center East (FRC-E) in Cherry Point, NC. He and his team of 200+- engineers, scientists, technicians, and support staff - provide sustainment and maintenance engineering support for all assigned Navy and Marine Corps aircraft, equipment, and support systems. The multiple type/model/series Navy and Marine Corps aircraft (with a special focus on helicopters, vertical lift and short take-off and landing aircraft) assigned to their team include the AV-8B, C-130, F-35B, H-1, H-53, H-60, MQ-8, RQ-21, VH-92, and the V-22. They cover a broad range of aviation specialty areas within disciplines that include quality engineering, materials & process engineering, composites, structures, rotors, engines, drives, starters, mechanical systems, avionics, automated test equipment, test program set development, and support equipment.

Mr. Meno earned a BS in Chemical Engineering from Virginia Tech in 1994. Upon graduation, he was employed at FRC-East (which was then Naval Aviation Depot Cherry Point) and served as a materials engineer responsible for organic coating removal, corrosion control, and aircraft and hot section engine cleaning technologies deployed in the FRC production environment. He then moved into management serving as the analytical chemistry and composites supervisor. He moved into command leadership soon after, first as the Engineering Sciences Department Head, followed by the Air Vehicle Department Head. He spent a year serving as the command's first Senior Civilian (Executive Director), after which he was designated as the Head of Research and Engineering in 2015 before his current position which began in the fall of 2019.

Made in the USA
Monee, IL
20 September 2023

43048356R00050